How Evolution Produced The Religious Mind

Volume I

How Evolution Produced The Religious Mind

Volume I

George Vernyn

About the Author

George Vernyn was born in Boston, Massachusetts, in 1944. He took BS and MS degrees at the College of Engineering of the University of California in Berkeley. He also studied Darwin under Professor Richard M. Eakin. He served in the United States Peace Corps as a teacher of science, music and computer programming in Africa. He then forged a lifelong career in electronics engineering, developing automatic factory machinery for the worldwide electronics industry, and designing subsystems for medical therapy equipment. He consulted to two university molecular biology laboratories, creating computerized systems that controlled automated electrophysiology experiments, processed fluorescence video microscopy data, extracted features from images, compensated for known artifacts, and rendered the data presentable. He is at his best when combining observations from separate disciplines, and making conjectures about previously unrecognized cause-effect relationships.

How Evolution Produced The Religious Mind Volume I.

George Vernyn

Copyright © 2015 George Vernyn, All Rights Reserved. No part of this book may be reproduced in any form by any means available in the past, present or future without the express permission of the author or his heirs and assigns. Additional restrictions may apply to quotations from other sources that have been inserted into the text of this book with permission from their original copyright holders.

First edition, January 10, 2015

ISBN-13:   978-1502558527

ISBN-10:   1502558521

Dedicated to our hardworking teachers, good and bad:

to the former, who urged us to think on our own,

and to the latter, who left us no other choice.

> "War does not determine who is right - only who is left."
> *Often attributed to Bertrand Russell*

Contents

1. Introductory Summary 1

How did evolution produce the religious mind? 3
Religion makes a man a better fighter 12

2. Innate Capacities in Humans And Other Animals 17

Man versus Animals, and Nurture versus Nature 17
Generalities about Capacities 19
A Time Before Religion 22
Will There Ever Be Religious Computers? 26
The Human Capacity for Forming Coalitions 31
A Quintet of Capacities 36
The Human Capacity for Mortal Combat 37
Not an Absolute Advantage, but a Differential One 41
How Much Code is Needed? How Much Hardware? 43
Butterfly Migrations 44
Nematode Pleasures 47
Gross Behavior Modification by Viral Infection 48
Gross Behavior Modification by Fungal Infection 50
Gross Behavior Modification by Protozoal Infection 50
A Small Code Can Do A Lot Of Work 51
Darwin is a Member of the Club 52
Nature Plus Nurture: Capacities Support Memes 54
Evolutionary Echoes in Present Day Humans 58
Instincts That We Do Not Possess 63
The Trolley Problem Exposed 68

3. The Capacity for Religion as a Heritable Instinct 75

About NoMa 75
Continuing the Heritability Argument 81
A Conjecture by Darwin 82

vii

1 of 8: The brains of religious and nonreligious people are observably different. 87
2 of 8: About 80% of the human population are religious, and about 20% are not. 98
3 of 8: *Bona fide* conversion from either side to the other is rare. 102
4 of 8: Religiosity and atheism tend to run in families. 104
5 of 8: The capacity established itself before the African diaspora. 105
6 of 8: Most People Grasp the Notion of The "God-Shaped Hole" 107
7 of 8: We have in hand a workable evolutionary explanation 108
8 of 8: We abjectly lack a competing explanation 109

4. Evolution Fails To Remove the Capacity for Religion 111

How Expensive Is Religion? 112
What Pays Religion's Costs? 120

5. Coevolution with the Capacity for Mortal Combat 123

Religion Makes a Man a Better Fighter. 123
The Mutual Binding of Separate Capacities 128
Coevolution with the Capacity for Mortal Combat 133
Prepping For Battle 136
Immediate Aftermath 138
The Splitting of Consciousness 140
Summary of Chapter 5 142

6. Additional Accelerants 143

The Capacity for Genocide 146
The Capacity for Kleptogamy 153
The Speed of it All 156
What About Today's Atheists and Pacifists? 157

Summary of Chapter 6 159

7. Writer's Apology 161

Not an Argument For Or Against Religion 161
A Gory Mess, and No Exit – Sorry About That 162
Shocked and Disappointed? 164
My Proofs are Weak 165
How Can Anyone Assert That This is Good? 169
A Bit Repetitious 172
This Story is Very Nihilistic 173
No, This Isn't Group Selection 174
Is Any Follow-Up Planned? 175
The Dissent of Man 177

Bibliography 179
Internet and Periodical Resources 203

Acknowledgments

I owe debts of gratitude to a large number of individuals and coalitions who have played a role in the development of these ideas and in the preparation of the book. You know who you are. I thank you. Some have expressed a need for privacy. The need to treat one and all with equal respect drives me to leave this page blank.

> "Alas! There exists an order of minds so skeptical that they deny the possibility of any fact as soon as it diverges from the commonplace. It is not for them that I write."
> - Andrè Gide

1. Introductory Summary

This is not an argument for or against religion. It's not about any particular religion, philosophy, or class of religions. What it IS about is Man's biological capacity to have a religion in the first place. It's a biology book. Any objective thinker, religious or not, who is possessed of a healthy curiosity and a basic understanding of the sciences could have conceived and tested the ideas I have written here. It is my hope that you will find here a fully rational, evolution-based explanation of the neural mechanisms that support human religiosity, and how they came to be installed in our genome and brains. I will do my best to keep the work free of value judgments such as "good" and "bad," and just report my findings at face value.

I will use the word "capacity" a lot to describe a neural mechanism that supports a particular activity of the organism. The human brain contains many thousands of capacities. One in particular (and there will be four more added to it as we move into the story) interests us: such a mechanism – a neural structure that facilitates the thoughts and actions of a religion – must be part of the brain in order for religion itself to be present. A creature can only do those things it is capable of. If a human capacity for religion didn't exist, the brain would not be able to carry out religious thoughts and actions: human religion as we know it could not exist. I will further explain that some humans come into the world equipped with this capacity and others do not.

The capacity, like all other parts of the nervous system, is implemented as an arrangement of neurons and synapses. The code for erecting and installing it during development and maturation of the body must therefore be part of our genome. Being heritable, the capacity must be shaped by the action of the law of natural selection. It follows directly that the capacity for religion exists as a consequence of our evolutionary history, I shall speculate on the details of that history.

I address no complex questions of theology or philosophy, such as have been addressed again and again in millions of words through thousands of years. There is little I could add to all that. I only explore one big question here. It is one of biological science, that is, this question of evolution:

How did evolution produce the religious mind?

The work began with recognition that the question itself (a) exists, and (b) is worth asking. About 25 years ago I found myself puzzling over the odd fact that some humans are religious, and some are not. We have all noticed this fact, haven't we. Why the difference? Is it due to nature, or to nurture? A common view, dating back at least as far as the 16th Century Catholic missionary Francis Xavier, is that the difference is all in the upbringing: some people are properly trained in youth, and some are not.

But this wasn't a good enough argument for me. There are too many exceptions. A great many adult atheists are people who received a full course of childhood religious training, and subsequently walked away from it. *Vice versa*, many adult believers of my acquaintance had made their religious discoveries late in adolescence, or had even become "born-again" as adults, not having received any formal childhood religious training at all. A difference in youthful training, in my view, was not sufficient to explain the difference in adult religiosity.

For any kind of training to succeed, I reasoned, the trainee must possess an underlying capacity to accept the training and to incorporate its thoughts and routines into daily life. For any given kind of training, some potential trainees will have the capacity and others will not. This fact applies to many known kinds of training. The "Three R's" come to mind in this context, it being well known that aptitudes for reading, writing and arithmetic vary considerably from one student to the next. Could religion be a fourth "R" like the first three? It is similar in that its full development relies on the success of a training process, and that the training process fails where the aptitude is lacking. I began to feel that differences in Nature, not just differences in Nurture, must be deeply involved in establishing this difference between the religiously-oriented part of the population and the non-religiously oriented part.

Although I introduce this topic in the apparent vein of "there are two kinds of people..." I recognize that many readers are suspicious of arguments that are built on brittle dichotomies. I know this intimately, being that sort of reader myself. So I ask a temporary indulgence on your part. I will smooth it out later.

I will commit a considerable part of this book to showing that purpose-specific neural mechanisms exist in humans as they do in other animals. There are a great many such mechanisms. Some, like those for breathing, for swallowing, for making eye contact, for crying, or for finding a nipple and sucking it, appear to be fully functional at birth, and start right up without any need for training: how, indeed, could you teach a newborn to look you in the eye? Other capacities, *e.g.* that for language, develop into full functionality gradually over the course of maturation, and provide platforms on which more elaborate structures can be built by training and experience.

I will assert that an inborn capacity for religion exists, and that it is of the latter sort, a neural mechanism not in itself religious, and not ready to go at birth, but essential, absolutely necessary, for the acquisition of a religion.

The capacity for religion, though very common, is not present throughout the entire human population. It appears to be a majority phenomenon rather than a unanimity phenomenon. As noted above, some people utterly lack it.

Those who have the inborn capacity will have the ability to receive the training, to make the frequently-reported ecstatic leap of faith, to accept the beliefs and rituals of their religion, and to comprehend the myths and stories. They will be able to accept and internalize all as a fount of blessing and truth. They will have passion, and they will go on to teach others. The capacity somehow shields them from being intellectually bothered by religion's inconsistencies and dead ends. The contradictions between religion's supernatural truths and the facts of science that are present in other forms of knowledge will rarely be confronted head on.

Those who lack the inborn capacity, on the other hand, are never be able to fully embrace a religion no matter how skillfully they are taught or how diligently they study. They cheerfully learn the words and the moves, *i.e.* the Creed, the Collects, the scriptural traditions and all the other elements, but at some point begin to question it all. The logical portions of their minds take the supernatural information apart piece by piece, and are never again able to put it back together. What they would have to believe does not match what they know. Unlike the true believer, they confront the disjunctions in what they are being taught, and seek logical resolution.

Religious training is pleasant to them nonetheless. They enjoy the meetings and the friendships. They agree with the moral teachings, which fit perfectly with their innate sense of probity and fair play. They thrill to the hair-raising rumbles, snarls and resonances of the organ and the bells. They love the art and architecture. They love the echoes, the glass, and the colors in the big hall. They figure out the chord progressions of the mighty hymns. They love the synchrony of singing in the assembled multitude. They learn all the poetic recitations and incantations. They read the lesson. At life's passages, they dutifully perform the prescribed rites. They show up to class on time, take all the tests and pass them. They may meet lifelong friends and spouses in this context.

But here's the thing: through this whole process, the emotions of the congenital nonbeliever are quite different from those of the believer. Lacking the capacity to give their minds over to religion, the religion-incapable see the elements of religious training in a different way. The learning is not difficult, but in their ponderings, they store up myriad bothersome questions. No matter where they look or whom they ask, they never get satisfactory answers. They can shelve the questions for weeks or months at a time, but cannot make them disappear. They cannot reconcile their beliefs to their knowledge. Even some of the simplest religious vocabulary words (for example "pray," "worship," "devil," "holy," "sacred," "faith" and so forth) fail to acquire any dependable meanings.

At some point in the training, or even long after it is completed, cracks appear in what once looked like a mighty, permanent structure. The shelved questions come down off the shelf. Gritty, pesky doubts pile up in dunes and drifts. They encroach on the space that others would reserve for the mysterious and the dogmatic. At some point, the Creed morphs into a true-false test: one's innate sense of honesty forces him or her to answer "false" to most of its elements. Such people are, in a way, mentally disabled. They are god-blind. Some folks are no good at mathematics: these folks are no good at religion. They are the congenital atheists, and there are more of them than you might think there are.

Now to smooth out that dichotomy problem: the capacity for religion is a spectrum phenomenon, ranging from "A" (atheist) to "Z" (zealot). It's not smoothly linear. There is a kink, a blank spot on the dial, a clear divide between the complete nonreligiosity of "A" and the point where the spectrum of actual religiosity begins. Given this explanation, the notion should be clearer than it was where I left it a couple of paragraphs ago. The people who have the capacity have it to a greater or lesser degree. Those who lack it, though, lack all of it.

Training (to a first approximation) offers no possibility of creating the capacity in one who lacks it, or of removing it from one who has it. There is no real moving of anyone from either side of the divide to the other. The inborn capacity is there, to some degree, or it is not there at all. Exceptions surely exist, if only as statistical outliers: not all accounts of them are fully credible.

What constitutes this capacity, this postulated neural structure? How many neurons are there, and how are they interconnected? We don't know this yet. I'll assert that the mechanism for religion fits easily within the available space, and that it isn't intrinsically costly from an evolutionary point of view. It is fairly simple, and doesn't require a vast number of neurons. Likewise, the genetic code elements that control its architecture don't take up a large number of DNA base pairs.

This idea – a genetically modulated inborn capacity for religion – is directly opposed to the "Blank Slate" view that I had been taught in freshman Psych in the 60's. I didn't trust that view. I turned my attention away from psychology and toward engineering. The engineering discipline provided a set of organizing principles that I found far more dependable. Half a century later, after completing a full engineering career, I found my interest in psychology revived. Times had changed.

[Pinker1997] and various other books had put the subject of psychology on an evolutionary footing. I found ideas similar to Pinker's in [Darwin1859], which was published a century and a half before. Could I apply this mode of thinking to the capacity for religion, *i.e.* could I study the evolution of the capacity for religion?

I have always tested ideas by attempting to explain them to myself in writing. If I could get the description to work for me, I reasoned, I might then share it with others. Private writing allows one to consolidate fleeting ideas, and fit them together into an integral frame. It worked for me, so I decided to share it. If you, too, find the question interesting, perhaps you will find good material to discuss in what I have written here.

The innate capacity for religion is widespread, but not universal in the population. I postulate that it is heritable. Being heritable, it is subject to the law of evolution by natural selection. If it evolved according to the law of natural selection, it must also have been adaptive in some way at some time, strengthening its possessors relative to others. But here is the paradox. Religion as humans practice it is very expensive, demanding time and material resources as well as a degree of intellectual containment. Wouldn't the law of evolution have acted to see all the resources available to humanity invested directly into specific activities that directly foster population growth? Examples of such activities include time, material and energy devoted to the care, protection, feeding and training of healthier, more numerous offspring. Such activities also include time and development of toolmaking skill for hunting, gathering, and processing materials for food, clothing, medicine and shelter. It includes putting land and labor to productive use. It includes weapons for defense and for combat.

Following this reasoning, evolution should have removed the capacity for religion. Instead, evolution has preserved it. It is this paradox that interests me, and I choose to resolve it. At some point in our evolutionary career, the capacity for religion must have provided some adaptive power. What is – or was – that adaptive power? What was its mode of action? This question cries out to be explored, so let's explore it. How does religion's adaptive power manifest itself? How does evolution favor it? What's the process? Is there a context? Is some other instinct cooperating with it, so that they operate together as a coevolutionary pair? Where does the capacity exert its evolutionary value? Where is the win? Or as a philosopher with a background in Latin might ask, "*cui bono?*"

There may be more than one mode of action. I will discuss just one here, one that is sufficient to explain the phenomenon. In a nutshell, it is this:

Religion makes a man a better fighter

This simple statement turns out to have a library of consequences. It picks up several corollaries along the way. I report just a few of those in these pages. If the consequences can be verified, perhaps the whole structure of the resolution will make sense: I therefore offer, to the world at large, a rational, evolution-based explanation for the fact that evolution favors the capacity for religion in our species. Let's see how it works.

Death by mortal combat has been part of human and pre-human life for millions of years. In mortal combat, the man with religion has a competitive edge over the totally rational man. The passion and the risk-tolerance of the religious man are greater. He is more quickly decisive, as well. The rational defender stops to ask "can't we talk about this problem?" Because mortal combat is, after all, mortal combat, and it's quick, those may be his last words. The religious man's advantage may be slight, but its action is quick, lethal, and permanent. Mortal combat is an ancient part of the human landscape, and occurs often. The principle of compound interest applies, making a slight advantage ultimately decisive in the long game. An inherited property that brings an advantage in combat must become prevalent over time.

Mortal combat also requires a neural capacity. It must. We are the only mammal species that practices mortal combat to such a total degree. We must therefore be the only species with the capacity. Thus, two capacities – one for mortal combat, and one for religion – work together. They cooperate, they merge, and they have coevolved. Either one, by itself, might be seen as an evolutionarily costly, but together, they have the effect of rapidly attenuating ("annihilating" might be a better word) the gene pool of their religionless defending neighbors, and replacing it wholesale with their own more powerful one. Thus, the adaptive power of this pair of inherited capacities can be described like this: those who have the capacities will get to reproduce and those who lack either or both won't get to reproduce.

Once this gets started, there's no stopping it. It could sweep across a continent in just a few generations.

I'll go into the details in the body of the book. I felt I needed to tip my hand here, rather than lose your interest… and to the end of keeping your interest, I promise even worse news to come, toward the end of the book.

Evolution stories often use euphemisms such as "survival," "die out," "compete," "adapt," and "fitness." Indeed competition and dying-out occur in this story as well. Traditional evolution stories rarely mention murder and abduction as means of accelerating the process of selective dying-out and competing. Mine will, though.

After a while, I began to present this solution informally to many friends and good thinkers over lunches and eventually in an OLLI seminar. It was often met by expressions of shock and amazement, doubt, disbelief, and disgust. One thing it has not yet been met by, though, is a workable evolution-based counterproposal.

In my reading, I kept hoping to find that someone had already written my book for me, but none of you had. In 2014, I attained the Biblically ordained maximum age of threescore and ten (Psalm 90:10). I couldn't predict that anyone else would ever write this book, so it became my existential duty to do it, and quickly. Here it is. I have done it.

A fair warning: Some readers may find this explanation disagreeable, even ghastly. Me too. I shall provide an appropriate apology at the end. For now, guard your mind a little from the gory mayhem that you are about to read. Consider it a study in the evolution of five capacities in one particular animal species. Hold at a distance the notion that you are a member of that species. And, if you have a better explanation as to how evolution produced the human capacity for religion, please publish it. I look forward to reading it, if you can get it done in time.

My purpose here is to explore the circumstances of the feature's arrival and of its preservation under the relentless action of the law of evolution by natural selection. An excellent sourcebook on that subject is [Darwin1859]. Very readable, and he seems to have thought of everything, so if you need a refresher on that basic principle, that's a good place to start.

"Although organisms sometimes appear to be pursuing fitness on behalf of their genes, in reality they are executing the evolved circuit logic built into their neural programs, whether this corresponds to current fitness maximization or not. Organisms are adaptation executers, not fitness pursuers." – David M. Buss, ed.

2. Innate Capacities in Humans And Other Animals

Man versus Animals, and Nurture versus Nature

To introduce this topic gently, I need to mention that I, like many of you, grew up in a crazy time long ago, when some of our elders taught us that "Man" was not an "animal," but instead, something special. Our psychology professors in the 1960's went so far as to teach us that "Man has no instincts." They cited great authorities. We believed them, for a while, but they were wrong, of course. Elaborate explanations notwithstanding, we found out later what we had known already – that Man does in fact have instincts, and not just a few but a great many. We are born with lots of automatic machinery in our brains. Some of it is ready to go at birth, and some of it takes months or years to develop. Each of us gets a slightly different set of instincts. The instinct array, like the genotype that produced it, will necessarily be somewhat different from one person to another.

Human instincts had been known to Darwin at least as long ago as 1859, e.g. from [Darwin1859]:

> "In the distant future I see open fields for far more important researches. Psychology will be based on a new foundation, that of the necessary acquirement of each mental power and capacity by gradation. Light will be thrown on the origin of Man and his history."

In my copy, those words are on page 759; they are omitted from some other editions. My use of the word "capacity" in this book is derived from Darwin's meaning. Now, a century and a half later, is Darwin's "distant future." Researchers around the world are constructing "evolutionary psychology," which is presumably just what Darwin meant when he proposed that "psychology will be based on a new foundation." As for the light that "will be thrown on the origin of Man and his history," I hope that this work contributes a few photons to it. I will show that in certain circumstances there are certain kinds of evolved instincts that are extremely powerful when they act in concert with other, previously evolved, instincts, and that "gradation" is sometimes more sudden than Darwin imagined. Some kinds of change may find ways to sweep rapidly through an entire population.

Enough, then, of the gentle introduction and the debunking of deceased psych professors: one way or another, we are ready to move into the core of the matter.

Generalities about Capacities

We are animals. We have located the chronological positions of the last common ancestors that we share with many of the other creatures in the tree of life. We have instincts. We have innate capacities. They are neural constructs. The designs of instincts are carried from one generation to the next through the genes. They are subject to the action of the law of evolution. Some, like the capacity for religion, are variably expressed in the genome, and will appear in one genotype but not in another. In other words, some of us get that capacity, and others do not.

Some capacities are installed and operating at birth, but others take time to develop as the child's body attains its various stages of maturity. Some are ancient, having evolved a very long time ago, possibly in a predecessor species, and have been carried forward into our current genome. Others are more recent additions, having sprung up after our line diverged from its preceding line. The passage through the birth canal, or in the case of Caesarean delivery, around it, is but one milestone on a lifelong path of neural development. We say that human instincts are "innate," *i.e.* we are "born" with them inside us, whether they are turned on and active at birth, or whether they are developmentally scheduled to appear – and join forces with the environment – later in life. In short, the fact that an instinct is "innate" does not mean it is up and running at birth. The fact that some external stimulus – *i.e.* "nurture" – is required to fully instantiate a capacity does not mean it is not "innate." The instantiation requires nurture, but the capacity itself is pure nature, innate. Anything that is "innate," *i.e.* part of the genome, is subject to the law of natural selection. If a capacity, *e.g.* that for language, is innate, then that capacity is subject to the law of natural selection. It will grow or shrink in the population over succeeding generations, according to its adaptive power, *i.e.* its ability to increase fecundity. Its adaptive power may vary, as time passes and situations change.

How can all these capacities fit in the brain, and how can the codes that construct them fit into sperm and egg? Imagine for a moment the structure of an instinct's hardware in terms of the number of neurons that need to be created by mitosis, and the number of interconnections each must grow to meet up with other neurons.

Estimates of human mental hardware vary considerably, but even as approximations, large numbers suggest a story of time-consuming growth during development. The adult human brain has around 100 billion neurons in it. Each average neuron connects to 1000 others, resulting in about 100 trillion (10^{14}) interconnections. My favorite quip about how long things take (variously attributed to Woody Allen, Kurt Vonnegut, Ray Cummings and others) is "Time is Nature's way of making sure everything doesn't happen all at once." It's difficult to estimate how long a biological system could take to build 10^{14} individual, precisely custom-tailored synapses, each going from a particular spot on one particular neuron to a particular spot on another particular neuron. A fiber of some sort starting from one neuron to connect to another must set out to identify its target among the other 100 billion candidates and grow in that direction. There are "signpost cells" along the way that tell the growing fiber which way to turn as it grows through the colossal maze that is the brain. Any parent, any schoolteacher, indeed any person who retains childhood memories, can tell you that brain development does not happen overnight.

Many of our most interesting mental capacities are sockets for receiving and holding memes [Dawkins2006], [Dennett2006], [Dennett1995]. These take many years of child development to become active, and still more to absorb and cement their memes.

A Time Before Religion

To focus on this subject, I will assert that there was a time in geologic history before the capacity for religion existed on our planet. During that time, since there was no capacity for religion, there was no religion. This is an odd and vaguely disconcerting thought to many people today, the capacity being so well established, and religion memes being so thoroughly pervasive in the human population. Nonetheless, please stay with me as I forge onward with the proposition. The capacity for religion was once absent and is now present. It stands to reason that it arrived somewhere between then and now. Not only has it arrived, but it has also become very deeply rooted. It's set up to stay a long time.

It should be clear that the thought-stuff of religion manifests itself largely in the brain. I'd go so far as to say it can only exist in an animal with a brain. But don't ALL animals have brains or one kind or another? Prior to the Cambrian Explosion, variously dated at half a billion years ago, there were no multicellular animal life forms at all, hence no nervous systems, and neither need for brains nor material from which to make them. So religion could not have arrived on Earth any earlier than that.

Today, half a billion years later, the ineluctable algorithm of evolution has been grinding away all the while, and more than a million different multicellular animal species have come to exist. All animals, of course, are equipped with nervous systems, with which they sense the state of the world and decide what to do about it. Nervous systems are brains. What else would we call them? Other writers have aptly observed that if there's a brain, there's a mind. Is the capacity for religion present in the genomes and minds of all of these animals? Hardly: in fact, religion is vanishingly rare among animal species on Earth. Of about 6,000 total mammal species, only *Homo sapiens* shows any signs at all of having religion. In addition to the mammals, moreover, there are some 1.2 million known non-mammalian animal species. Science's total count of animal species is growing all the time: previously unidentified species get discovered, described, and named to the scientific community – to the tune of about 10,000 per year.

As we will discuss later, only about 80% of humans have the capacity for religion. These are the only creatures on Earth who have the innate neural capacity for religion in their brains. Our species is the only one known that exhibits signs of religion, such as grave goods, written symbols, ceremonial structures, songs, taboos, gatherings, fancy dress, *etc.* and even at that, it's limited to only 80% of us. The total penetration of the capacity for religion into the varied species of the Animal Kingdom, is thus about one in 1.5 million.

The human capacity for religion was once a mutation, an innovation in a population of nonbelievers. It was once new, just as surely as the opposable thumb the forward-facing eyes, or the erect bipedal posture had their origins: these too did not exist at some prior time, and are now standard equipment. Somehow, they work better than what went before.

The capacity for religion had to have been brought about by a mutation, a change in the DNA that could be passed on to descendants. The neural capacity for religion, once it arrived, spread through the population in a way similar to that of the opposable thumb and the upright stance, *i.e.* somehow, it worked better than what went before. The capacity for religion must have conferred some physical competitive advantage on those who possessed it, allowing them to "outcompete" their fellows. It made us better at something. We have already suggested what that something was (mortal combat), and we shall continue to explore that throughout these chapters.

For now, imagine again the time before religion, a time in the deep past when there was no religion at all. Can you do it? 80% of today's humans will have some trouble with that, whether they realize it or not. The other 20% of today's humans, who lack the capacity to be religious, easily imagine a time without religion. They actually live their private lives in that time. They would like nothing more than simply to live that life all the time. Today, of course, the 20% who lack the capacity warily observe and constantly adjust, due to the unavoidable and often judgmental presence of the other 80%.

Friends I have discussed this with often challenge me. "How do we know that other animals don't have religion? How can WE know what ANIMALS think?" they ask. "Maybe they DO have religion. Even rocks might have religion." I agree with part of that: we don't know what other animal species think, or for that matter what minerals or vegetables think. We barely know what our own species thinks. But who will take on the task of demonstrating that other animal species have religion? The burden is not mine. O you doubters, show us your evidence of rattlesnake religion, or of caterpillar catechisms! Even if you find one or two species, you will not make much of a dent in my ratio, and no difference at all in the general thesis.

I'm comfortable asserting that our pre-religious animal predecessors lived their lives entirely without religion, as do presently our 5,999 fellow mammal species, 1.2 million non-mammals, and about 20% of current humans. If my count is off by one or two, I don't mind. This observation – that the capacity for religion is extremely rare on our planet – is one of the things that prompted my work.

Will There Ever Be Religious Computers?

An instructive metaphor for the nature/nurture issue can be drawn from the world of computer operating systems. Application programs ("memes?") are written to run on computers with operating systems. They are written in such a way as to rely on services (capacities) from the underlying operating system software. The application program makes calls to supporting routines in the operating system, for example to send text and pictures to a display, to store information on a disk, or retrieve information from the disk, to check the time of day, detect mouseclicks and keystrokes and so forth. In our neural metaphor, the operating system is a set of capacities, and the application program is a meme that makes use of whatever capacities it needs.

Not all operating systems provide the same identical sets of services, and not all brains contain the same set of capacities. Consider an ancient computer whose entire user interface consisted of a punch card reader for input and a text line printer for output. Some of us are old enough to remember using such machines to perform what we regarded as useful work at the time. The operating system for this old computer might have been a hardwired patch panel. What we now consider an "operating system" originated with the IBM 704 in the middle 1950's. It did not need to provide support routines for interfacing keyboards or visual displays other than at the operator's console, since those devices were uncommon. It would not be possible for that ancient operating system to support a word processing program.

Wait a few more years, though, and time-shared operating systems that supported "remote teleprinter terminals" became available. The operating systems for timeshared systems provided services for reading keystrokes from, and sending characters to, the teleprinters. Applications programs could use these services to interact with human users in what we called "real time." These innovations with respect to teleprinters facilitated the development of application programs to perform online editing of text files. Online editing quickly evolved into word processing, email and so forth, features that are universally familiar today. Next, user interfaces evolved into mouse-and-graphic systems, touch screens and spoken voices. What comes next, we don't yet know.

It is important to remember that punch-card based systems, though they were as capable as any of today's machines at calculating and tabulating, could not support word processing or email, let alone a Web browser. They could not respond to mouse clicks, finger touches or voice commands – important aspects of user interfaces in today's consumer products. Yet computers without sophisticated user interfaces did not disappear from the Earth. They are ever more abundant in our world today – they are the embedded processors in our cars, in our industrial products, and in our thermostats, sprinkler controllers, burglar alarms, entertainment systems, toasters, refrigerators, slow-cookers and so forth. They have operating systems and application programs, tailored to the needs of the devices that they control.

To close off this metaphor, consider the capacity for religion as a neural innovation – a new operating system service akin to the one for the teleprinter – a new tool for the application programmer to try out and use and stand back and observe what happens. At its outset, nobody can predict everything the user community will make of a new service, or how far it can go. At first, the novel capacity for religion immediately provided essential support to the application programs that we now call religious memes. Nobody could predict everything the user community would make of the capacity, or how far it could go, because the memes themselves were yet to come (this suggests the need for a capacity for creativity, but I will save that discussion for a later book).

An operating system – *i.e.* a brain built from DNA instructions – that does not provide the religion capacity will not support a religious meme, any more than a punch-card computer operating system will support word processing software. That brain's owner will be unable to run a religion meme (think "application program"), but can run all the older programs just as well as before. Aside from being incompetent at religion, the owner of such a brain will be perfectly normal, subject to all of the usual variations and the usual bell curves.

My career workmates and I (a full spectrum, including both true believers and frank atheists) designed, built and installed computers and computerized automatic systems over several decades. We often imagined similarities between our electronic creations and actual living creatures. Our creations were like a sort of silicon-based life that needed us carbon-based animals to serve as primal symbionts. It was harmless banter, of course, but I can't resist dispensing one relevant anecdote. Several years ago, I read a magazine article in which a great researcher (regrettably, I have forgotten his or her name) announced that he or she was about to construct an electronic machine that would exhibit a property called "consciousness." Strange. I brought the magazine article in to work, and revealed it during lunch with co-workers. What did they have to say? One proposed that the machine would need to be baptized right away. Discussion followed until another friend, without dropping a stitch, said that a baptism made sense to him as long as it would be done by total immersion!

There are now many researchers in the field of "machine consciousness." A Web search on that topic just gave me 18 million hits. I wouldn't bet on getting a religious computer anytime soon, though. Unless one regards the human brain as a "computer." And many do. A Web search for "computational theory of mind" delivers nearly three million hits. The way the computationalists view humanity, we already have religious computers: we ARE religious computers.

I opened this part of the discussion with talk of the capacity for religion, but only because my main purpose is to explain capacities and especially that capacity. It is not, however, the oldest of the five capacities that we'll take on. Prior to the biochemical innovation that became the DNA instruction for building our neural capacity for religion, our species (or one of its predecessor species) already had plenty of other instincts and capacities. It kept all those older instincts when religion joined the party; but with the addition of the capacity for religion, a vast – and sudden – transformation took place in the destiny of mankind.

The Human Capacity for Forming Coalitions

Religion is not the only neural capacity that interests us. It acts in concert with other instincts. I will discuss a quintet of capacities that work together. One of them is coalition-forming: a very much older, genetically determined, neural capacity is the part of our genome that wires into our brains the capacity to form coalitions. Today, we smoothly and skillfully form coalitions of many kinds, varying immensely in size, scope, and purpose, but all structurally similar. Each of us "belongs" to several coalitions at once, and effortlessly moves between the context of one and the context of another.

We live in a milieu of families, bands, tribes, clans, schoolrooms, schools, scout troops, sports teams, neighborhoods, villages, towns, cities, counties, states, nations, treaty organizations, social and political philosophies, support groups, kennel clubs, orchestras, service clubs, fundraising groups, political parties, clothing preferences, choruses, castes, parenting organizations, states, nations, military units, churches, religious sects, alumni associations, workplaces, job teams, unions, customers of a particular supplier, listeners to particular radio stations, readers of particular magazines, prison gangs, rows, columns, floors, buildings, and so forth.

Each of our coalitions is, in our point of view, in some respect "better than" other similarly constituted coalitions. It is OUR group. We regard our group as the special, on-top, ur-group of its kind. We see those "others," however closely related by interest and kind, as interlopers, opponents, or just "competitors." An older coalition may be regarded as musty and obsolete, and obligated to make improvements or else make way for our new, more modern and effective coalition. A new but related coalition may be regarded as a gang of heretics worthy of suppression. Under some circumstances, a coalition of "others" can be seen as "mortal enemies," as will be developed in a further chapter.

We set up contests, elections, athletic games and so forth to adjudicate or otherwise resolve power struggles between one coalition and another. We usually follow agreed rules in these confrontations. If our gang comes in second or worse, we resolve to make adjustments before the next contest. We analyze our defeats. Did our product have advantages that we forgot to deploy? Was the other guy's salesman better than ours? Can we hire him away? Can we bribe the judge a little better? Did some scandal occur that tainted our coalition and robbed us for an instant of our collective will? Do we have a weak member, a Jonah, whom we should throw to the whale? Have we been cursed by a witch? Can we wreak something similar on our opponent, or at least fool him into believing that he or she has been bewitched?

I can imagine, through a fog of time and approximation, varieties of what the coalitional instinct might have looked like at its inception, and how fast its installation period progressed. I know the capacity is old, because without it, we would be a solitary species, not a social one. In one scenario, I imagine a uniform distribution of non-coalitional proto-humans living solitary lives, spaced equidistant from one another across the unforgiving savannahs of Africa, getting together periodically at the lek [Ardrey1961] to find mates. As each observes the others, he or she scouts the field, and makes assessments of the strengths and weaknesses of other individuals of its species, still viewing each as competitor, and likely to be hostile once the annual meeting is over.

An errant cosmic ray modifies a tiny stretch of DNA in one germ cell or one gamete, and an instinct for coalitional behavior arises in the children that descend from that mutation. Soon a "team" appears on the savannah, working as a social unit. Maybe just two brothers at first, or a parent and children. The individuals in the team tolerate each other's presence, solve problems cooperatively, encircle and gang up on solitary opponents. They share joint memories around the fire, and exchange stories of their travels. They attempt either to recruit or to ruthlessly destroy whatever individual competitor they happen to have encircled. If they fight, and fight to the death (mortal combat will be a common theme in the book, as it is in world history), it's pretty obvious whose genes are going to win that contest. What contest? Well, imagine yourself as one of the loners (or should that be spelled "losers?") You lack the mental equipment that would allow you to cooperate with your neighbor to establish a common defense, form a more perfect union, *etc.* (phrases borrowed from [Constitution]).

You win the fabled "Darwin Award" - good-bye! A social species is born, and a solitary species disappears. Other than that, you were both pretty much the same.

The ancient biochemical genetic event that created the neural capacity for coalition-building set the stage for many future genetic innovations in humans and other primate species. Perhaps the species in which this coalitional capacity one day burst forth was not even identifiable as proto-human, but was some pre-simian pre-primate ur-ancestor. The monkeys and the apes (note to older readers: apes, chimps and hominins are now all called "hominids" in current convention: humanlike hominids, whether extant and extinct, are now called "hominins.") are said to have diverged from a last common ancestor around 35 million years ago. The hominin line is said to have branched from the chimpanzee line, out of a common mother who is thought to have lived about 6 to 9 million years ago. Today's monkeys, chimps and hominins are all coalitional, *i.e.* social, like us. Was the last common ancestor of monkeys and apes also social? If so, the coalitional trait may go back beyond 35 million years.

It's also possible that the gene package for the coalitional capacity arose more than once, of course. In that case the human capacity for coalitionality may be younger.

Coalitionality is a pretty good neural innovation, though. It would likely be conserved by evolution wherever it arose. I leave that question for others to refine. For my purposes, I only need to assert that the capacity for coalitionality already existed when the capacity for religion arose.

The coalition instinct is so basic, and so all-pervasive in today's ape species (of which we are one, don't forget), that it's generally assumed, *i.e.* almost not worth mentioning. Yet, I felt I needed to mention it, because the capacity to form coalitions is foundational to all of the arguments that I'm going to make. Everything I'm going to describe here depends on our social, coalitional, competitive nature. I assert that the capacity to form coalitions (and to compete against other similarly constituted coalitions) is an innate instinct, and thus subject to Darwinian natural selection. Another way to say this is that the law of evolution acted to produce the coalitional mind. The capacity for coalition-forming actually results from the simple, relentless, recursive action of that law.

A Quintet of Capacities

There are other human instincts that the law of evolution has chanced to stack on top of the capacity for coalition. I will describe four others, to form the quintet. You are invited to make your own speculations on many more. The quintet of capacities consists of those for coalitionality, mortal combat, religion, genocide, and kleptogamy.

The Human Capacity for Mortal Combat

One particular inherited instinct worth mentioning at this point in the story is the human capacity for mortal combat. This capacity, like that for forming coalitions, is ancient. Humans have had it at least through all recorded history, and likely for many centuries, or even tens of millennia before. It's even observed in the fossil record, in the form of certain kinds of skeletal injuries both healed and unhealed. Fossil ulnar bones broken in "parry fractures" are generally regarded as evidence of a serious fight, with the break having occurred in a defensive stance. A fossil ulna with an unhealed parry fracture indicates the defeat (*i.e.* death) of the combatant, while a healed break suggests survival, or possibly a team victory.

Fossil skulls with holes broken into them also speak volumes about ancient combat events: some anthropologists go into detail about the positions of the holes, and what each one means (was the assailant right-handed, for example.) Related species that lived a million or more years ago, and which may be ancestral to ours, evidently engaged in intra-species killing much as our current species does today. Mortal combat possibly arose and built itself on top of the capacity for forming coalitions. Many prehistoric conflicts between one coalition (*i.e.* one band, clan or tribe) and another resulted in, and were perhaps temporarily resolved by, mortal combat.

The details of the arrival of the mortal-combat capacity in the human genome are not known, but often credibly speculated. One vivid speculation: in their film "2001 a Space Odyssey" Arthur C. Clarke and Stanley Kubrick depict their imagination of the instant of its arrival. With great drama, they act it out. Two tribes (coalitions) of apelike protohumans routinely challenge each other at the watering hole, engaging each other only in harmless howling, breast-beating, scowling and stomping. After the "Monolith" appears, though, signaling that something has changed, the leader of one tribe discovers that a long bone (possibly a bovine femur) can be a powerful destructive device – a weapon.

First he practices, crushing bovine skulls and ribs. Then, the intra-species confrontation takes place. He challenges the leader of the other tribe to single combat, and quickly clubs him to death with the long bone. The rest of his tribe, now also armed with large bones as clubs, then join in the violence, pounding the last bits of life out of the flaccid body of the former, but now fallen, champion. The scene ends with the victorious leader (he is given a name: "Moon Watcher.") flinging the long bone high in the air, as the surviving members of the defeated tribe slink away. The tumbling femur morphs from a weapon into a space ship bound for the Moon, where pretty soon we will meet ... another Monolith. Are we its primal symbionts? Is it ours? N.B. this movie also has a conscious computer in it...

The origin of the capacity for mortal combat may or may not predate the split between the chimpanzees and the rest of the apes, variously estimated at 6 to 9 million years ago. Our species and its line of descent are included in the phrase "the rest of the apes." The "may or may not" uncertainty is rooted in the observation that one of two extant strains of chimps occasionally engages in intra-species mortal combat, and the other does not.

Did the peaceful Bonobo (*Pan paniscus*, formerly known as "pygmy chimpanzee") line branch off before the main chimp line split from the rest of the apes? If the last common ancestor between warlike humans and warlike chimps was equipped with the capacity for mortal combat, then the bonobos must have somehow arranged a way to shed that capacity, while the rest of the chimpanzees retained it. I have trouble imagining a way in which evolution would shed a capacity of such huge competitive evolutionary power, but you might have an idea to contribute. Or, perhaps the bonobos split off a little earlier, before the mortal combat capacity arose on the rest-of-the-apes side of the split.

It's also possible that the mortal combat capacity arose independently in chimps and humans after the chimp/human split. Like the capacity for coalition-forming, the capacity for mortal combat is pretty strong at biasing the outcome of a contest, and evolution would be likely to conserve it. Gorillas, gibbons, and other apes appear to lack the capacity for intra-species mortal combat. But humans possess it to an astonishing – some say revolting – degree, further confusing the logic of the last-common-ancestor argument.

The details of that descent I also leave to other researchers. What cannot be denied is that humans are to a large extent instinctually wired to form coalitions, and also to engage in mortal combat against members of other coalitions of their own species. One need only leaf through an encyclopedia of world history such as [Langer1952] to see that through almost all documented time there has been at least one known war raging somewhere on the planet. There are fifteen or twenty going on right now as I write this. They are in Africa,

Asia, South Asia, the Middle East, the Arabian Peninsula, and Meso-America. Try as we might, we have not yet discovered a reliable way to block the action of the human capacity for mortal combat, or even to blunt the constant human desire to engage in it.

How did evolution conserve the genes for the mortal-combat capacity? At first glance, we'd say that it would be unlikely. The capacity costs vast numbers of human lives. Aren't lives valuable for the success and growth of the species? Evolution keeps only one figure of merit on its scorecard, the rate of population growth. How could a genetically programmed loss of human lives be considered an adaptation, a fitness for survival, an evolutionary success story?

Not an Absolute Advantage, but a Differential One

The competitive advantage of the mortal-combat capacity is differential, not absolute. The traditional "absolute" view of gradually evolving fitness deals with heritable traits that make a species more successful as a whole, integral species against other species. The less-frequently-discussed "differential" view deals with traits that enable some members of a species to vanquish other members of its own species in combat. It works like this: the successful killer survives to breed again, while the failed, dead victim does not. He has bred his last. Perhaps, if young enough, he has not bred at all, and leaves no descendants. The failed fighter, whose capacity for mortal combat is missing or in some way attenuated, is gone, and his genes are gone from the gene pool. He might have been mentally preparing to negotiate a peace and work things out with his opponent, but unfortunately, he died before he could bring the subject up. Thus, when a victim loses the mortal battle, and dies – because he lacks the capacity for mortal combat – his genetic contribution to the future of the species is differentially attenuated. That of the winner is differentially amplified. This process results in the advance of the combat capacity through the descendants, fight by bloody fight.

Also consider that the mortal combatants – whether winners or losers – tend to be male. Females, not males, are the limiting resource for the human species' reproductive capacity. Males are pretty much expendable from evolution's point of view. Only small numbers of males are necessary in most species. A species such as ours could lose almost all of its male members in battle, yet still manage to keep all of its female members pregnant and producing offspring.

It is also a tradition that great fighters are celebrated in song and story within their bands. They are lionized and are granted increased reproductive opportunities. It may be more than a tradition – we might be wired for that, too. We shall discuss the fate of the women in the losing band in a later chapter.

This idea of a differential advantage will come up again when we get into the discussion of the capacity for religion.

How Much Code is Needed? How Much Hardware?

At this point you may be asking how masses of instincts for behaviors as ethereally complex and intricate as coalition-forming and mortal combat can be encoded in the mere 22,000 or so genes of the human genome, especially when there are so many other things to be encoded as well. It might not seem enough, at first glance.

Hold on, though. Perhaps the codes for these behaviors, and the neural structures that put them into action, are simpler than we imagine them to be. A given instinct, such as that for coalitional behavior, might be accomplished by an unexpectedly small number of synapses among an unexpectedly small number of neurons. The simple architecture might be programmed by an unexpectedly small number of base pairs in an inconspicuously short patch on a strand of DNA.

What suggests to us that the hardware for an instinct might be cheap? There are many well-studied creatures that instinctively accomplish acts of seemingly impossible complexity with small numbers of neurons. Here are some examples:

Butterfly Migrations

Some species of butterflies migrate vast distances, processing all their information through a brain the size of a grain of sand. And no matter how small that brain, controlling the migration is only one of its many duties. The migration module of the insect's brain must contain stable data about the route. It must contain and process the algorithms of navigation. The insect must cope successfully with capricious winds. Its tiny brain successfully calculates the vector cross products needed to compensate for the Coriolis acceleration, as the butterfly moves north and south along the rotating globe's longitude lines. The butterflies are social, so may possess a version of the aforementioned coalitional capacity. They might school, as many fish species do. To make the matter more extreme, remember that the lifecycles of such butterflies are shorter than the time it takes to complete the migration. The butterflies must reproduce along the route. The adults die, leaving the next leg of the migration to the next generation. Each newly-hatched animal, born in mid-migration, is headed for a place it has never seen before, and will never return to the place of its birth.

To put this another way, the butterfly's DNA – a collection of 16,866 genes – seems to contain all the information necessary to construct the animal's body, including the neurons, which it will assemble into a working brain with sensory, logical and motor processes. Part of it must be dedicated to providing instructions for building the computing networks that the brain will use when it needs to find the route, with its attendant waypoints and destinations, and to recover from meteorological disturbances.

Humans are not yet fully agreed as to how the butterflies' migration and navigation programs work, or what information inputs they rely on, but work is underway. Theories abound, involving magnetic fields, and light of various wavelengths and polarizations. A front-runner seems to be the "time-compensated sun compass."

A few meta-questions struggle to be asked: how did evolution produce this system we are inquiring about? What were the capacities of the Monarch butterfly's predecessor species? Is there a non-migratory species of the Monarch, and in how many places does its genotype differ from that of the migratory one? Could we find something relevant in a related species, for example the fruit fly? The fruit fly is the most common laboratory animal in the world, and its genetics are known in great detail. The fly serves as a model and testbed for many interesting experimental gene manipulations. Could fruit fly data throw any light on the quantitative question about butterfly behavior? How distant is their last common ancestor?

One reason I bring the butterfly problem up is to show that small bits of DNA and a small number of neurons can orchestrate prodigious and highly specific, detailed, long-range, trans-generation behavioral tasks. I speculate we can use the information to refine our guess as to how simple the chemistry that controls our own capacity for religion might be. It's got to be simpler than what the butterfly uses to handle its migrations. Another use of the butterfly story is to stimulate curiosity in the power of evolution (a very simple algorithm) to construct systems of high apparent complexity. A third reason for bringing up the butterfly is suggested by the remark about the fruit fly model and its last common ancestor with the butterfly. This is to remind us that we tend to use models, either actual or mental, to refine our guesses as to how things came to be as they are.

For deeper discussion see the "Monarch Butterfly" article on Wikipedia. Now let's go on to some other estimates.

Nematode Pleasures

The genome and nervous system of the roundworm *Caenorhabditis elegans* have been extensively studied in laboratories all over the world (a Web search for *Caenorhabditis elegans* returns over 3 million hits). Its genome has about 19,000 genes (the human genome has only a few thousand more, the Monarch butterfly a couple of thousand fewer). Its nervous system contains only 302 neurons, yet these few manage a fully functional animal that feeds, digests, excretes, moves, responds to stimuli, remembers things, and seeks food. Most interesting with respect to the concentration argument, eight of the 302 neurons are said to be devoted to "pleasure." Of course it is not easy for our 100-billion-neuron mind to understand exactly what "pleasure" means to a 302-neuron mind, but you can look this up in a learned publication, for example [Linden2011]. For the purposes of the book you are reading right now, the point is that neurons are extremely capable cells with lots of functionality. Biology can do a lot of computing with very limited hardware.

Gross Behavior Modification by Viral Infection

Along that line of thinking, briefly consider the baculoviruses, one of which preys on the tree caterpillars of the gypsy moth. This virus, acting on the gypsy moth caterpillar, is the cause of "tree-top disease" (German "*Wipfelkrankheit*") in moth-infested trees. The virus doesn't infect the tree directly, though: it only infects the caterpillars. Infection by the virus brings about a radical change in the behavior of the host caterpillar. The mechanism has been identified: the virus modifies the production of one particular kind of molecule, a molting hormone, in the caterpillar. A normal, uninfected caterpillar feeds in the treetop during the night, but returns to lower levels during the day to hide in the soil or in cracks in the bark.

The caterpillar normally molts five times before pupating. When it's ready to molt, its molting hormone is secreted, and the caterpillar stops feeding long enough to molt. An infected caterpillar, though, stays in the treetop, feeding all day as well as all night. The virus has interfered with its ability to produce the hormone that would have made it stop feeding and molt. The infected caterpillar, lacking the hormonal cue, cannot stop feeding, and it cannot molt. At the right moment, it attaches itself high in the branches, dies and liquefies, reducing itself to a sack of maturing virus particles.

The baculovirus particles rain down on the healthy caterpillars below, and infect them. The caterpillar's behaviors of climbing and feeding, and anchoring at the treetop to die are adaptive to the virus: in raining down from a higher altitude, dispersion is improved. The parasite has hijacked the brain of its host to help it increase its reproductive effectiveness. Baculoviruses, according to a Wikipedia article, exist in many forms, and infect over 600 hosts, causing subtle changes in the behavior of each. Their relationship to the gypsy moth must be deeply ancient.

Baculoviruses effect complicated behavioral changes in their hosts, typically by affecting the production of a hormone specific to the relationship. How many genes does the baculovirus maintain in its little genome? 29 at the core, plus a few more specifically tailored to the particular host animal. Simple mechanism, complex lifecycle. As with other items in this list, there's a meta-question: how did evolution produce the baculoviruses?

Gross Behavior Modification by Fungal Infection

A natural history similar to the *Wipfelkrankheit* cycle has also been observed with respect to the fungus *Ophiocordyceps unilateralis* that infects carpenter ants. The infected ants become zombies in the service of the fungus. The excellent Wikipedia article on the ant-fungus parasitism notes that it was discovered in 1849 by Darwin's friend and voluminous correspondent Alfred Russel Wallace.

Gross Behavior Modification by Protozoal Infection

Another well-known parasitic relationship is that of *Toxoplasma gondii*, which infects mammals such as mice and rats. It fills their brains with asexually-reproduced cysts, but can only carry out sexual reproduction inside the gut of a cat. Infected mice, it is said, lose their fear of cats and are more easily caught and eaten than uninfected mice. Once the cat eats the infected mouse and, along with it, the cysts of its parasites, the parasites reproduce sexually in the cat's gut. Oocysts containing new *T. g.* zygotes are discharged into the environment in the cat's feces. Mice that eat such cat feces become infected, and the cycle continues. Admittedly the mouse's brain is far larger and more complex than the brains of caterpillars and ants, but the parasite itself, whose genetic code orchestrates the behavioral change, is very small indeed – a single cell protozoan. See [McAuliffe2012] for an account of *toxoplasma* parasitism in humans.

A Small Code Can Do A Lot Of Work

I include all these examples – butterflies, roundworms, caterpillars, ants, and so forth in order to suggest that it doesn't take very many neurons, or very much DNA, to establish or modify a behavior pattern that seems complex to us humans. For more examples, see [Zimmer2014].

Not to beat this a-little-does-a-lot thing to death, but one more thing, maybe metaphorical, maybe more than metaphorical: good mathematicians from Leibniz to Mandelbrot have been fascinated by the ways in which complex phenomena can arise from small collections of short, simple laws, formulas or algorithms. I think of such people as members of a club, the "simple rules, complex results" math club. The rest of us all think a complex phenomenon bespeaks a large and complex set of rules.

Leibniz, though, working in the 17th and early 18th Centuries, considered that our universe has to be "better" than all other possible universes because its amazing complexity appears to be produced and governed by a few simple physical laws such as those of gravity and momentum. Mandelbrot, working in the 20th and 21st Centuries, defined and popularized fractal systems. His most famous fractal is the range of beautiful, intricate pictures that unfold from the recursive iteration of a single line of code in a computer.

People who enjoy a small, tight, formula that explodes into enormous complex beauty absolutely love the Mandelbrot fractal. See examples at [Peitgen1986] or see the Wikipedia article on "Mandelbrot Set." The fractal picture is so complex that it's often referred to as "chaos." In fact, it is totally predictable and repeatable. It isn't random. It arrives at its precisely defined state of complexity from a single line of code in a computer, confined in an iterative loop. If you run the program a second time, you'll get the same picture again.

Darwin is a Member of the Club

It may seem odd, to some, that I regard Charles Darwin as a mathematical innovator of the same kind as Leibniz and Mandelbrot. To me, he's a charter member of the "simple rules, complex results" club. He qualifies for admission by virtue of his discovery and careful explanation of a simple rule, the law of "natural selection" ("evolution," for short.) He is particularly admirable for his patient and thorough exploration [Darwin1859] of the consequences of evolution's action over countless iterations. His formula, like that of Mandelbrot, is an irreducibly simple algorithm. Iterated, after the fashion of Mandelbrot, it explains in one stroke life's simple mechanism for fully occupying its environment. Its consequence, over countless iterations, is the vast radiation of life itself into millions of different forms, each with its own astonishing complexity – built entirely from simplicity.

Evolution, once Darwin explained it, rapidly became the central organizing principle of all biology. To borrow a phrase from Newton, Darwinian evolution is the "system of the world." It's not just a good idea: it's the law.

A moment to re-explain and hopefully cement the notion that the theory of evolution is important in the thesis of this book.

- Humans are animals;
- Animals get to be the way they are by evolving under the law of natural selection;
- An animal's genome governs its qualifications to survive in its niche;
- Life forms can become endowed with complex behaviors that arise from small numbers of neurons and small numbers of genes;

- Humans are equipped with many capacities, including one for religion;

- The capacities, including that for religion, are innate;

- Being innate, capacities are subject to the law of natural selection;

- Being heritable under sexual reproduction, some capacities may exist as phenotypes in only part of the population;

- In some past time, the capacity for religion improved human fitness for survival.

Nature Plus Nurture: Capacities Support Memes

It remains for us to discuss the details of the process that catapulted the capacity for religion into majority representation in the human population. Is the capacity for religion evolutionarily useful today? Was it more useful at a previous time, say in prehistory, and has it simply lingered through the ages to today? In what way did it – or does it – improve our reproductive fitness? Does it combine synergistically with other innate capacities?

Previous authors on human religion (they are legion) have mostly overlooked the role played by biological evolution. For one thing, the acquisition or removal of a religion has been generally viewed a matter of training, *i.e.* nurture not nature. It is clearly more than that. Religion, like language, is not something that can be acquired by a formless, gray slab of memory, the proverbial, inchoate blank slate. Clearly, training is necessary for one's acquisition of a particular religion (or a particular language, or membership in a particular coalition, and so forth), but the training is completely lost on a person who lacks the innate capacity to be trained.

This is a key point, and I will mention it more than once as we go along. Wherever a meme such as a language, religion or country is to be hosted in an individual's brain, an innate capacity is required to support the meme. There can be no religion in an individual who lacks the capacity for religion, no language in an individual who lacks the capacity for language, and no patriotism in an individual who lacks the capacity for coalition. All of these neural capacities, and all others as well, result from the action of evolution.

The overriding purpose of my writing is to re-inject evolution into the discussion of human religion. Any discussion of human neural development that doesn't include a discussion of evolution is incomplete.

One upshot of all this discussion of how simple things cause complex reactions is simply that we should not be overly concerned about the problem of paucity of genes. I am pretty sure we have enough to encode all of our instincts as well as all of our other physical manifestations.

Getting back to human instincts, consider a few basic instincts that are at work immediately upon human birth, *i.e.* with a baby's first gasping breaths of air. The new infant can cry. When presented with its mother's skin, it begins rooting around, seeking a nipple. On finding the nipple, it begins sucking. No training precedes these activities.

In the next few days or so, the infant can look you in the eye. As an engineer who spent several years developing so-called machine vision systems, I can tell you that looking someone in the eye is not an easy piece of computation. Yet the otherwise seriously incompetent newborn baby can find your eyes and lock in on them. It knows when you are looking back, and when you are not. Not too long after that, the baby can recognize individual faces – a computationally intensive task that thousands of engineers have been struggling with for at least the 75 years that computers have been available. And soon after it starts recognizing your face, the infant can look at your face when you are NOT looking at its eyes, and figure out what you are looking at: it can follow your gaze.

We know these complex behaviors are instinctive. If they had been installed by "nurture" rather than "nature," how did we go about teaching them to the infant? We did not. The machinery was already there, and it was started up by a breath of air. It is a clear case of nature, not nurture. For that matter, there are mirroring instincts in the adult, parental brain – we love the little baby, we celebrate its accomplishments, we are troubled by its cries of hunger, discomfort and loneliness. Its eyes draw ours in. Do these adult instincts operate when we are newborn? No, but they will be ready to go when we have a newborn of our own to love in the future. Are they part of our training? No. They appear right on schedule, as soon as we have a baby of our own. That is the point when they become evolutionarily useful. As soon as we become parents, we become susceptible to having the new baby trigger those instincts, along with powerful emotions of attachment and devotion.

So those are a few little newborn inborn capacities, along with adult inborn capacities, each innately tuned to the other. All are pretty complex from a computational point of view. Yet we know they are not taught, not cultured. They are nature, not nurture. We don't know how many genes are involved, but we can be sure that it's not too many for our genome to carry with it.

Evolutionary Echoes in Present Day Humans

These human capacities are wondrously complicated, and we accept their heritability without much of a question. Yet, we still ask, is the newborn brain capable of supporting something so complicated as a capacity for religion? Not while the brain is still figuring out what a face is, but later on, resoundingly yes. We don't know if the capacity for religion is simple or complex. I suspect the capacity is built using elements of other capacities, and is in itself fairly simple. To get a feel for something more complex than finding a face, let's explore some aspects of child's play – from an evolutionary point of view, child's play is serious business.

A year passes after birth, and balance and gait grow in importance, as greater sensing, computing and actuation capabilities arrive. In walking and running, clumsiness diminishes, and to a certain extent, actually vanishes.

Airborne Boy With Grandmother, Photo © 2011 G.V.

Jumping down from a height (after climbing it) is a computationally intensive and highly dynamic operation. How complex? 75 years of rapid discovery in electronic computing and development of control system theory by tens of thousands of human engineers have not yet produced a robot that can do this; not one that WANTS to do it, and especially not one that decides WHEN to do it, or is CONSCIOUS of doing it.

Could a child's neural system learn all these desires, operations, trajectories and decision points from an outside teacher? In other words, is the part of the child's brain that carries out these actions anything resembling a blank slate? Resoundingly, no. There's more nature than nurture at work here. If teaching was involved in the acquisition of this routine, then who taught the child, and how? The child's grandmother, looking on skeptically in the background, definitely did not teach the child how to do this. Grandma can't even do it herself, and has no desire to re-learn the trick. She vividly remembers doing it when she was his age, but no longer feels any need to do it herself. She didn't teach him WHEN to do it, HOW to do it, or reason with him about WHY to do it, either. She probably would prefer that he DIDN'T do it, but knows she cannot stop him.

The child repeatedly climbs the concrete caterpillar and jumps off to the rubber padding below. He lands on his feet every time, and barks out a little shout of joy. He runs back around to the tail end, climbs the caterpillar again, and keeps doing it over and over, keeping it up long after his doting grandparents have lost interest in the process, and have stopped "reinforcing" the behavior. Perhaps they have even begun to try de-inforcing it instead. But he keeps going, until HE decides he's done with it. So there is another question worth considering: What does he get out of it, or as Dennett might ask, "*cui bono*" [Dennett2006]? Perhaps the child is driven onward by an evolutionary need to build his glorious sense of the body's power and balance, and solidify it by repetition.

Here's an expansion on that idea, in the form of an evolutionary echo, a legacy instinct. The human species, or possibly a predecessor species, is or was the beneficiary of whatever is going on here. It isn't as important today as it was several tens of millions of years ago, but this run-climb-jump practice is still there, built into the human genome. Let's imagine that what's going on here is the solidification, by repetition, of the computations, coordinations and synchronizations that are necessary to carry out the climb, the jump and the landing at will. It looks pretty safe today, but ... was this kind of repetitive jumping practice a risky move, way back in the distant past? Was it risky enough to be an evolutionarily selective game, one that removed some of the ancestral child's less competent age-mates from the gene pool? Is it an echo of a childhood developmental activity in an arboreal population, millions of years ago? I note that this young survivor has his arms spread high and wide, perhaps for balance, or perhaps in preparation for grasping a branch?

I imagine young, ancestral, primate children raining down from the forest canopy to their deaths, leaving in the population aloft only those who had the innate capacity to build – and exercise – the computational power required for jumping safely from branch to branch. These tree-borne survivors lived, grew up, found mates, and passed their heritable neural capacities (including those that foster jumping safely from branch to branch) on to their descendants. The fallen youngsters, lacking the right hardware, missed their landings, failed to keep their balance, misjudged their distances, lost their grips, *etc.* and departed the ancestral gene pool with a whistle and a bang. Still other departees may have not had the instinctive urge to practice the routine as children, and never carried out the repetition needed to solidify their jumping-down instincts: these, too, would have been less likely to attain reproductive maturity.

For their mortal troubles, the losers received only the posthumous imaginary "Darwin Award." The winners – our very distant ancestors – lived, matured, mated and reproduced. The winners passed their genes on down the line to us. As modern humans, we no longer find it evolutionarily necessary to swing and jump among tree branches, but we retain the fossilized urge to climb and jump at a certain stage in our childhood development. Implausible? Not everyone gets the preferred genes, and I'm not the only person who makes up stories like this: according to

http://www.darwinawards.com/ in 2012,

"... A South Carolina USA man took a "swan dive" out of the gene pool as he jumped between cruise ship levels. ..."

I freely admit that the child in the photograph may have seen other children playing on this concrete caterpillar, and just thought it looked like lots of fun. And even this is not an anti-evolutionary notion. Isn't fun, too, evolutionarily inspired? Isn't "fun" one of those instinctually-provided states that helps us decide what to do next? Doesn't the 302-neuron brain of *Caenorhabditis elegans* have eight neurons devoted to "pleasure?" How many does *Homo sapiens* have?

Instincts That We Do Not Possess

I'd venture to say that, before this little boy actually jumped for the first time, he had carefully made a mental model of the whole trajectory. He assessed the risks of slipping, and of falling. He estimated the force of the landing impact. He did all this in his mind, millisecond by simulated millisecond. He knew he could perform it without getting hurt.

Ironically, he was, at that time, still not ready to ride a carousel, one of which goes around merrily just up the hill above the caterpillar jump. He had been introduced to carousels and studied them carefully, but he didn't yet know whether or not he would survive a ride on one. A carousel to you and me is a simple contrivance of unbreakable wheels, cranks, lubricants, rollers, gears, and painted animal sculptures.

Adults and older children consider the carousel perfectly risk-free. The carousel, though, was not to this younger child's liking. He could not be persuaded to take a ride, at least at that age.

The carousel moved on its own, unlike the reliably stationary concrete caterpillar. This boy had examined several carousels, even unto crawling under one of them once to have a look while it was stopped. Being unable to see the wheels in the undercarriage in the dark, he might have thought its platform was somehow levitating, an observation that would have been hard to reconcile with his other understandings of gravity and mass. He was curious, but he had not yet figured out how it worked, or how its whole cycle played out. He had seen lots of other children riding it, but still regarded it as a suspect, and possibly dangerous contraption. Maybe they knew something he didn't. It had unreliable geometry, or some other repellent characteristic. He had no mental model of it, yet.

And maybe there's an evolutionary explanation of that, as well. Consider the child's arboreal forbears, and terrestrial ones as well. There were no carousels in their lives. There was no evolutionarily significant need to learn how to survive an encounter with rotating machinery. Accordingly, there was no need to evolve a mental model of a carousel, nor was there any need to make a mental rehearsal of what might happen to a person who went on a carousel ride. Evolution never had a chance to select for survival a child who comprehended carousels over one who did not. Today's human child, therefore, has neither an innate understanding of carousels, nor any drive to practice coping with them. Far fetched? Feel free to propose your own evolutionary explanation of this child's behavior and nonbehavior.

> Steven Pinker [Pinker1997] p.319 notes:
>
>> "... three- to four-month-old infants see objects, remember them and expect them to obey the laws of continuity, cohesion and contact as they move. ... Three-month-olds can barely orient, see, touch and reach, let alone manipulate, walk, talk and understand. They could not have learned anything by the standard techniques of interaction, feedback and language. Nonetheless, they are sagely understanding a stable and lawful world."

The statement appears to apply as well to this three- to four-year-old boy on the playground as it does to Pinker's three- to four-month-old infant groping for a favorite toy. Clearly a complex set of inherited capacities, not a blank slate, is what's doing the job. The child's information processing – the watching, the mental modeling, the choosing and deciding, the firing of muscle stimuli and the integration of visual and inertial inputs – originates and executes fully within the child, triggered by only a few tiny cues of external input from the visual system. The child even seems wired to seek and receive those specific cues, another kind of meta-capacity. It wouldn't be too much of a stretch to say that some adult seems have been wired to design and build concrete caterpillars surrounded by bouncy rubber landing zones for the training and pleasure of little children.

This particular child, who does all the clambering and jumping so well, is too young to explain it or to have it explained to him. The grownups do not intellectually sit down and write his programs and do not verbally bootstrap his computer – we do it all chemically, through our DNA. He does all his work with hardly any external language. By the time he is old enough to install enough conversational language to discuss it, he may have forgotten how he felt about the carousel; he will probably find the jumping a bore, though it was, for a few days or weeks, the absolutely greatest, most ecstatic, pleasure in his entire life. It made his day, day after day. Perhaps this picture, if he sees it, will reawaken memories of the day when I took it, but it's unlikely he'll ever again jump off the caterpillar with as much gusto as he once did.

So it is with some kinds of development. Once this phase of maturation has done its job and all the relevant synapses are emplaced and tested, the scaffolding gets taken down. The activity need not be regularly revisited. Anyone remember how to crawl? It used to be all we did, when we were four or five months old. Older children rarely crawl, except as a joke, or maybe to get under an overhanging obstruction.

Toddlers on the edge of language vocalize in a babbling way, trying out all the sounds that a human might ever be able to make, keeping the ones that seem valuable, and pruning back the others. Babbling ceases when conversation starts. It is no longer needed. Interestingly, though, babbling may resume for a while when a young child is introduced to a second language. If the learning of the second language is begun too late, say, after the age of ten, the speaker will be stuck with the primary sound repertoire of the first language, and have great difficulty overcoming his "accent."

The Trolley Problem Exposed

This chapter would be incomplete without a short discussion of at least one well-studied mental process that, like the carousel problem, doesn't appear to have any instinctive operating system support. It is called the "Trolley Problem." A Web search for that phrase gives 19 million hits, so it's in wide circulation, and you are welcome to look it up and study its details and variations. It's based on a trolley car, usually driverless, and a number of people tied to the tracks, helpless as the car approaches, threatened with imminent death. It is often cited in studies of philosophy, ethics or moral choice, but rarely in studies of biological evolution. Here's a quick summary of two of the more popular scenarios.

In one scenario, you discover that five men are tied to a trolley car track. A driverless trolley car is speeding toward them. If you do nothing, they will be run over, and all five will die. In your hand, you discover a lever connected to a track switch that will divert the lethal vehicle to a sidetrack. Should you pull the lever? Sure! Do it! But wait, Oh No! There's a man tied to that sidetrack as well! Stress! Someone will die at my hands no matter which choice I take! Even knowing about this smaller horror, the majority opt to pull the lever. They kill one to save five.

In a different scenario, you are on a bridge over the track, which crosses below you. There is just a single track with five people tied to it. A driverless trolley car is fast approaching. There's no switch, no sidetrack, and no lever for you to pull. What can you do?!?! Aha! You are conveniently standing next to a fat man. You quickly realize that you can push the fat man over the rail in such a way that he will fall on the tracks, derail the trolley car and save the five who are tied to the track. The fat man will die, of course. Most people, oddly, opt to leave the fat man alone and let the five people die, though.

So each scenario offers a possible outcome in which only one dies and the other five live, and an alternate outcome in which five die and one lives. People who are given this exercise don't all make the same choices. They tend to make up outlandish stories to justify their choices. We seem to choose one in one scenario, but the other in another. Philosophers discuss at length how people go about evaluating the two scenarios differently.

Here is my biologist/engineer take on the Trolley Problem. Our wiring, or lack of it, makes this a difficult task. We have no evolution-generated wiring that would solve it automatically without ambiguity. Lacking such a "Trolley

Problem module" (neural equipment that's pre-wired for making a quick and sure decision), we are forced to think it through a piece at a time, and assess each step using our innate Golden-Rule-based valuation system. We must weigh each cost and benefit separately for ourselves and for the people we will save or kill. We can get stuck in a bind. It would be easier if we had a handy instinct that would tell us that the one-dead-for-five-alive was of equal cost in either scenario, but we apparently do not. We are forced to perform as stepwise reductionists, not as holistic grab-and-go intuitive responders.

We lack a Trolley Problem instinct, in my opinion, because our distant progenitors never encountered any regular antecedent for it (such as the one we have postulated for jumping down from a height.) Primitive humans had no trolleys, no tracks, no levers, no track switches, no bridges, and possibly not very many fat men. Sure, we might have had a couple of the elements, but the gestalt of the Trolley Problem never presented itself in antiquity. If our distant forbears were indeed arboreal, for example, they must have had instinctive ways of computing trajectories of thrown and falling objects; they'd have had a good chance of getting the fat monkey to crash onto the track in the right place at the right time. They certainly had the innate Golden Rule, too.

But taken as a whole problem, our distant arboreal forbears didn't have Trolley Problems, and didn't need to solve them. They were not presented with situations in which choices such as these mattered in an evolutionary sense, *i.e.* no individual would be selected to live or die based on the choice that she or he came up with. Accordingly, no instincts related to the Trolley Problem have ever evolved. There was no adaptive value in having an instant ethical response to a Trolley Problem scenario.

There probably still isn't any adaptive value in it, actually. I'm not saying the Trolley Problem has never happened among modern humans in "real life." Maybe it has happened, and maybe more often than I am aware of. But I read the news a lot. As far as I know, it hasn't happened to any of us. It's a pure simulation, a what-if whose results don't affect reproductive success. If it has happened in the real world, it has certainly not happened in a way that would result in the issuance of more fertile offspring from the person with one answer than from the person with the other answer. There is no Trolley Problem instinct within us today because there never has been a Trolley Problem that we had to solve to evolve.

Fortunately our always-creative philosophers have been able to invent such fascinating problems. They are interesting and possibly useful puzzles for stimulating thinking in a species that values thinking, but they do not bring to bear any survival skills, and do not suggest any differential advantages on the reproductive front. We take time to reason them out slowly and carefully. There's no automatic solver in our wiring that presses us forward, toward an immediate knee-jerk solution that we KNOW is right. If you are fortunate enough to find the time to visit the Wikipedia article on the Trolley Problem, you will observe several good thinkers tying themselves (and each other) into beautiful, creative, logical knots about it. That computational frenzy may be a mirror of the convoluted stepwise thinking that a normal human goes through when confronted with the Trolley Problem itself, being unable to invoke an instant, instinctual response,

If we had enough time and space we could discuss many more problems – REAL problems – that we pose to ourselves in the present day, and that our instincts are not adequate to solve. To name just two, how to run a corporation, and how to drive a car. These items were not factors in our prehistoric development, either in the trees or on the savannah. And it's not surprising that we're not very good at these activities. We compensate for our weaknesses with written handbooks, deep study, training, practice, licensing and governmental regulation. We also install road signs, and police the day-to-day operations. We set up insurance arrangements that help us pool the risks of our wrong decisions. The lack of reliable human instincts for driving cars leads to tens of thousands of needless deaths each year in crashes. This fact has stimulated more than one large corporation to undertake development work on automated cars - cars that will not need to be "driven" by a human at all. The self-driving cars will, in effect, obtain the instinctual capacities they need for "driving" themselves any year now; humans will, at long last, be relieved of the need to learn how to drive poorly. But this will happen only if the leaders of those corporations succeed in managing the engineering projects effectively (despite their utter lack of any instinctual capacity for managing large projects.)

In the next chapter, I will present thoughts that expand on the notion that religion is not a blank slate phenomenon universally accessible to all brains. Indeed, religion is a passenger on an underlying neural structure that has evolved. Since not all humans possess that underlying structure, not all humans are capable of instinctive religious thought and action. Some of us, lacking the capacity for religion, go through convoluted stepwise thought processes that a religion person might process instantly. I'm referring to mortal combat, of course, not to the Trolley Problem.

To review the points of this chapter:

- Humans are animals;
- Humans have instincts, just like any animals;
- Instincts, like other animal features, are subject to the law of evolution;
- Darwin wrote about this in 1859;
- Some instincts appear complex, but may be defined by simple codes;
- The capacity for religion is an inherited instinct;
- The capacity for religion was produced by evolutionary processes.
- When we are confronted with a problem for which we have no quick, instinctual solution mechanism, we fall back on a stepwise intellectual approach, which is slower.

We shall speculate on the details of those evolutionary processes in the following chapters.

> "Like nothing else, evolution really does provide an explanation for the existence of entities whose improbability would otherwise, for practical purposes, rule them out." – Richard Dawkins

3. The Capacity for Religion as a Heritable Instinct

In the previous chapter, we explored a range of instincts, some in humans and some in other animals. In this chapter, we will delve more deeply into the notion that the human capacity for religion is a heritable instinct.

About NoMa

> "The good thing about science is that it's true whether or not you believe in it." –Neil deGrasse Tyson

To get this chapter started on a stable footing, I need to explain that we are about to encounter a popular social and philosophical norm known as the doctrine of the "Non-Overlapping Magisteria," or "NoMa." We are also going to flout it. Readers who have already discarded this relic of the late 20th century are invited to skip ahead a couple of paragraphs while I write my own rejection of it.

The NoMa way of thinking reaches back to the Middle Ages and possibly further, under other names. Religion and science are seen as a duality – two separate and complementary bodies of process and information, each having its own "*magisterium*" or "teacher of doctrine." Religion helps the human address one set of questions, and science helps address a different set, so the story goes. Under NoMa, the sets are disjoint. In some explanations, science asks "how" and religion asks "why." I've always preferred to work in the "how" sphere, and will do so here, though for a reason unrelated to NoMa. I'm an engineer, wired, trained, and tested for "how" work. Why ask "why" if "how" will lead more rapidly to useful results? There may indeed be an answer to "why" evolution produced the religious mind, but you're unlikely to find it in these pages.

The NoMa truce proposes that religion and science should not overlap. Unfortunately, how this is to be accomplished has never been precisely explained. Each is to courteously stay out of the other's province? The NoMa truce allows science its free exercise in every subject not claimed by religion. Science, in effect acknowledging religion's primacy in all things, respects those claims, and acquiesces to the rule that some areas of human interest – notably religious ones – are off limits to scientific inquiry. What could possibly go wrong with that?

The ultimate consequence of some areas being off limits to science, *i.e.* of some subjects being outside of science and unavailable to its questions, speculations, probes and researches, is nothing less than the complete destruction of science. A major feature of science is its rejection of arbitrary borders. Science is our way of describing, discussing and testing universal natural laws. It requires us to inquire, to speculate, to measure, and to boldly suggest new discoveries. We must reconcile conclusions derived from one area with observations made in another area. This simple act merges those formerly disparate areas into one, and annihilates the boundaries that formerly divided them.

If science works anywhere, it works everywhere. If there's a place where it doesn't work, the enterprise is in doubt, and remains devalued until the discrepancy can be explained. There can be no exclusion zones, no reserved domains. Science presumes that all phenomena are exposed for the inspection and testing of universal laws of nature, some of which are yet to be discovered. Western science has had only about 400 years of reasonably free inquiry, during which our species has been busy discovering and expanding our understanding of nature's laws. It's just a short time. We know that we have not yet discovered them all. Our overriding assumption is that we have not yet come to the end to our ability to discover, propose and test such laws. We have surely not yet come to the end of our ability to use the known laws, either singly or in blended groups, to synthesize fresh explanations of what we see around us and how it came to be.

In the world of science, nothing is "outside of nature," or "above nature." Nature is all we have and all we are. There is nothing but nature. The very word "supernatural" is meaningless. In the world of religion, we are led away from studying nature's mechanisms, into trust and admiration of ineffable mystery. The habit of science is to convert mysteries into problems, and then convert problems into solutions.

If we as practitioners of science were to develop an understanding of something previously considered "outside" of the agreed upon realm of science, that thing would immediately be enveloped by it; our knowledge of nature would immediately grow by its inclusion. We would be forced to ask new questions about it, and to continue the great work.

The hare of religion, on waking from his nap at the fabled finish line, would then awake to find that he is no longer anywhere near that finish line. Scientific discoveries made while he slept have caused the finish line to move forward. The tortoise of science, who never understood the concept of a "finish line" in the first place, has gone plodding on ahead. The hare will always manage to catch up, take a shortcut, pass the tortoise, and draw a new finish line – and then settle down for a fresh nap. We should not worry for his future, or for that of the tortoise.

As *reductio ad absurdum*, imagine a spot in the map of the Universe (*i.e.* in "nature,") that has been arbitrarily rendered invisible to science by NoMa's *magisterium* of religion. Blanked out. Beyond the Pale. Not accessible to the lesser *magisterium* of science. What is science to do, if its randomly blundering, tortoise-walking, path of inquiry accidentally needs data from that spot? Get the data, that's what. The conclusion of any successful scientific inquiry is always something that needs to be bolted on to the common frame of already-known natural laws. Anything new that does not fit – or is undecidable for lack of data – demands adjustments of one kind or another until the new piece either fits the common frame and becomes part of the frame (*e.g.*

Kepler's and Newton's laws of motion), or gets shelved for later re-examination (*e.g.* perpetual motion, cold fusion).

If we are not allowed to look into the blanked-out spot, or to report what we saw there, how does the world know whether or not some vital information is occulted there? Is some piece of natural knowledge hidden there, some data point that could affect the outcome of our rational inquiry? If we imagine a trajectory whose beginning and end we know, but whose middle we do not, do we have a discovery with enough integrity to be worth reporting? To remove those hidden points from our inspection is to weaken the entire enterprise.

So, let's focus on the notion that a well known scientific law of nature – Darwinian evolution by natural selection – has operated over time in such a way as to produce the capacity for human minds to be religious, that is, for human brains to support the mental processes required for religious thought, will and action. Science, it seems, is investigating the very roots of human religion. It would be impossible for science to do this in compliance the doctrine of NoMa.

Lest the hackles of some readers begin to rise at this point, I will restate my opening remark here: "This is not an argument for or against religion." Every system of religious beliefs and disbeliefs is completely safe from attack by me. Instead, it is your scientific side that I wish to encourage. I wish to explain, in terms of basic biology and deep prehistory, how the human mind evolved the capacity to possess – and be possessed by – any of the thousands of religion memes that are at home on our planet. I am sure I can do this, and reasonably confident that some readers will find the explanation useful.

Continuing the Heritability Argument

This particular chapter continues the argument begun in the previous chapter, that the human capacity for religion is a heritable instinct. It is widely present in the population, but falls short of saturating it completely. What evidence suggests that the capacity for religion is a heritable instinct? It would be easy to answer that question if some researcher had identified a specific stretch of DNA, or even just a strong hint of a marker in the human genome – a chemical constellation – that exists in one state for people who are equipped with the capacity for religion, and in a different state for people who lack it. I hope someone is working on this, but as far as I know, it has not been done yet. It might not be easy, but we might luck out. The inheritance pattern might be first-order Mendelian – wouldn't that be a gift? If the capacity involves linkages of adjacent genes, such as are well known in fruit fly work, the inheritance pattern would be bent away from strict Mendelian ratios. Worse yet, the inheritance pattern might produce ratios that are close to Mendelian, but arrive at those ratios by non-Mendelian processes. Not being a geneticist, I can provide little more advice on the details of the mechanism that regulates the heredity of this capacity. It is enough for me at this point to argue that heredity is involved in passing the capacity down the line, and move on with the story. If heredity is involved, evolution is in charge of what happens next, regardless how the actual chemistry is organized. (Darwin, also not a geneticist, worked this way, to good effect.)

A Conjecture by Darwin

Darwin (I digress but remain in context) was keen on making predictions based on his evolution theory, even though he knew nothing of the underlying chemical mechanisms that enable it to work. In one case, he was presented with the Madagascar orchid *Angraecum sesquipedale*. This plant's flower has a nectary at the bottom of a very narrow spur that can be as much as 16 inches deep. The anthers that carry the plant's pollen are down there in the spur, too: how does this flower get pollinated? Few people understood how that could have evolved, since no pollinator insect had yet been discovered with a proboscis that long.

Darwin was ridiculed by some for suggesting that, hiding in the bushes, awaiting discovery, there must exist a species of pollinator insect, probably a hawkmoth, with an outlandish 16-inch proboscis. His publication on that subject came out in 1862, a mere three years after his 1859 masterwork. Darwin, who died in 1882, did not live to see the discovery of the pollinator moth that he predicted, but in time – 1903 – the moth was indeed found, collected and described. It was given the name *Xanthopan morganii praedicta*. The Latin adjective *praedicta* refers to the fact that Darwin, by conjecturing its existence, had predicted its eventual discovery. This episode is richly chronicled on the Internet, and anyone curious about it should at least look at the relevant Wikipedia articles. A short movie of this moth in action is part of the Cosmos television series by Neil deGrasse Tyson, in 2014 Episode Six, titled (appropriately for the hawkmoth) "Deeper, Deeper, Deeper Still."

Lacking (for the present) any concrete chemical evidence for the presence or absence of a religion capacity in a DNA molecule, we must rely instead on our holistic observations, and create conjectures that might explain how the capacity came about. (Note: the Latin roots of the word "conjecture" mean "to be thrown together." This is not a pejorative term.) If our ideas prove strong enough, we may arouse the interest of some competent geneticists and molecular biologists in designing the necessary experiments and carrying out the lab work. If the proposal involves government funding, there are sure to be interesting debates about the ethics and the politics that must be aligned. Hopefully, they will only retard progress for a short time.

What holistic cues will we explore in this chapter, cues that suggest that the human capacity for religion is a heritable and innate instinct? I will limit myself to eight, listed below.

- First, the brains of religious people and nonreligious people act in ways that are observably different.

- Second, about 80% of the human population are religious, and about 20% are not.

- Third, *bona fide* permanent conversion from either side to the other is rare.

- Fourth, religiosity and atheism, like many other mental tendencies, tend to run in families.

- Fifth, the capacity, like a great many other mental capacities, is found worldwide, which suggests that it established itself before the African diaspora.

- Sixth, most people intuitively grasp the notion of a "god-shaped hole," but not all people possess such a hole.

- Seventh, we have, in this book, a workable evolutionary explanation, in which the religion capacity's main adaptive characteristic is known: it provides instinctual decisiveness during mortal combat.

- Eighth, I don't know of a competing evolution-based explanation.

It might be useful at some point to attempt a definition for the term "religion." In this context, though, it might not be wise. Let's not get our hopes up. A Web search for "religion definition" produces nearly 200 million hits. Negotiating a definition on which we all agree could consume our valuable time to no good end, and lead us off the topic, never to return. Rather than have us distracted from the primary subject at hand – how evolution produced the religious mind – I'll accept that we may have different religions and different definitions of the word. Most religions have mythology, ritual and fellowship. Most have descriptions of supernatural beings and expectations of life after death. Some use chemical stimulants to enhance their rituals. Some use rhythm and music. Many maintain ancient books, which they call scriptures. Religions are coalitions. Religions have leaders, champions who "defend the faith." Adherents to each religion regard adherents of other religions as less enlightened than themselves. All rely on their adherents being capable of a sphere of thought in which some phenomena are constitutionally inaccessible to analysis by laws of physics.

There are surely exceptions, aren't there? You can probably think of one or two right now. So, rather than dwell further on the definition problem, I will direct you to others more expert than I to supply the occasional necessary detail or unrecognized generality. First, I borrow a deathless line from a former US Supreme Court Associate Justice, Potter Stewart (he was commenting on pornography, but I think it applies equally well to religion). Justice Stewart is reported to have said that it is "hard to define, but I know it when I see it."

Another valuable source is the philosopher Daniel C. Dennet, who does a very good job with the details — I recommend [Dennett2006] to all readers who enjoy watching a brilliant man do fine precision work with small, sharp tools. My takeaway generality from Dennett (though there are many others) is that religion is "hard to fake." Dennett actually borrows this phrase from yet another level of expert, William Irons (*q.v.*) Religion is indeed, above all else, hard to fake. It can be faked, but that's very difficult. It requires constant vigilance on the part of the faker. Anyone who has spent valuable time trying to fake a religion could tell you this. So could anyone who had once believed that he believed, but found out later, to his dismay and consternation, that his fellows were the true believers, and that he had been unintentionally faking it the whole time. True believers don't fake their religions, and don't have this problem.

In the aggregate, then, religion has hundreds of millions of definitions. It is hard to fake and hard to define, yet we know it when we see it. It has a lot of subsidiary properties, too. The main thing we must ask evolution about regarding religion's various properties is "are any of these properties adaptive, *i.e.* do any of them foster differentially increased fecundity?" Are their strengths reinforced by their undefinability and by the difficulty of faking them?

1 of 8: The brains of religious and nonreligious people are observably different.

How do we know that the capacity for religion makes the brains of religion-enabled different from those of religion-disabled people? Medical scanning technology is not yet ready to identify a physiological difference – and as noted earlier, that physiological difference may be exquisitely small.

Likewise, DNA correlation work may be possible and should be attempted as soon as the right grants can be obtained, but investigators will have to cope with uncertainties, chief among them being the candor of the individuals being tested. So, we shall have to proceed by examining behavior.

What, then, is the observable difference? The aspect that particularly concerns us here is a property of the mental hardware, not of any particular religious tradition. Aptitude for scientific thinking varies from person to person. People who have that aptitude can, by definition, be trained to think scientifically, whether or not they are also equipped with the capacity for religion. An engineer who has the capacity for religion can respond to an engineering challenge just as incisively as an atheist engineer. The capacity for religion neither amplifies nor diminishes one's ability to think scientifically about things outside one's own body.

But when it comes to thinking about the workings of one's own personal mind, there IS a difference, and it can be observed. No matter how ingrained its scientific side may be in rational, natural thinking, the mind of the religiously inclined person shrinks from making scientific inquiries into its own personal religious workings. It keeps its science and its religion separate, and uses them in separate ways, at separate times. It is as though the capacity for religion divides the mind in two – a rational side and a divine side – and builds a NoMa-like wall between them. People who lack the capacity for religion lack the divine side, of course, so don't possess this interesting, externally observable, division.

What right have I to assert this? First, I came to realize it, gradually over many years of personal interactions with friends who exhibited these properties, same as anyone else would. The religious engineer plies his rational trade all through the workday. He or she depends on his designs to work if the known laws of nature are dependable, which they are. Outside of worktime, he or she often prays for mystical guidance and special favors, which are unrelated to the laws of nature, and may even require that the laws be changed and bent away from their known functionalities. By and large, this person will avoid discussing this apparent contradiction. One physical brain contains two separate minds that are apparently not in communication with one another. When one is "on," the other is "off." Though the splitting of the brain's consciousness has not yet been observed structurally, the concept appears to explain the behavioral observations in a functional way. It kicked off a decade or three of personal reading and friend-testing. So, all you neuroanatomists, molecular biologists, bioinformaticists, geneticists, fMRI workers and PET-scan people, kindly get to work and see if you can find physical evidence of a structural difference between those with the capacity and those without.

I saw corroborating hints of the brain division while reading about "Pascal's Wager" (see the Wikipedia article). The deep and brilliant Blaise Pascal (1623-1662), while attempting to find a logical way to decide which way to "bet" on the existence of God, recognized that some people are unable to believe in a god. Yet he thought it very important for each person to make a conscious, rational decision whether to believe or not. In effect, Pascal's Wager is a bet that one will have a decent afterlife in exchange for mild sacrifices while one lives. Well, Dr. Pascal, nice try. If one lacks the capacity to believe in one's religious teachings, how does one make a rational choice to do it? Seems to me the only rational choice for an honest atheist is to disbelieve, and that isn't even really a choice. That there is an afterlife penalty in store for the unbeliever is equally unbelievable, though, so maybe things will work out OK for believer and nonbeliever alike. A major fallacy in the Wager problem is Pascal's assertion that the pleasure vectors for Heaven and Hell, (assuming they exist at all) are equal in magnitude and opposite in direction. This may or may not be the truth. No evidence is cited. If Heaven's not as wonderful as some say it is, and if Hell's not as horrible as some say it might be, then the wager's payoff is not particularly consequential. Pascal urged nonbelievers to at least understand their nonbelief and try to convince themselves to overcome their deficits.

Pascal is said by some (*e.g.* Edward T. Oakes, *q.v.*) to be "the first modern Christian." He had the mixed blessing to live and do his thinking in the 17th Century, a time of great intellectual progress, but also a time of the Thirty Years War, and two centuries before Darwin. I see him struggling to let his rational side examine his religious side, but failing. He is tormented in the attempt. Nonbelievers don't worry about such things.

I enjoyed a long and fertile career as an engineer, working side by side with hundreds of other people all committed to the same technical goals – identify the needs, invent, design, create, construct, mass-produce, deliver, install, teach and maintain sophisticated capital-equipment products. The machines were made of analog and digital electronics, software, electromechanical actuators, vacuum pumps and valves, servomotors, optical sensors and cameras. They had power levels from nanowatts to kilowatts, and frequencies of interest from DC to light.

Each machine was required to work reliably over long periods of time in all the countries of the world. Each was required to survive programming errors made by unknown users, and to report accurate measurements under a variety of conditions, within known tolerances. If something didn't work the way it was supposed to, we applied the scientific method to isolate the cause, intuit a corrective action, and test to see that the fix really solved the problem. The most interesting problems were the intermittents – some test would fail in apparently random ways, then start working right again all by itself. We could always fix those, too, because we KNEW that the cause of the bad behavior was not truly random. It always had a root cause that could be identified. We applied disciplined solving methods. In our company, we formalized a Japanese version of the method of Western science as the "Seven-Step Process."

In running that method, time after time, we all relied on our training in the laws of nature, *i.e.* the laws of physics, chemistry, thermodynamics, electromagnetism, fluid mechanics and the like. We depended on those laws to be universal and invariant. The laws simply had to work, everywhere, all the time. And they did. No exceptions. We'd send a machine to Costa Rica or Malaysia, and it would work just the way it did at home. We used scientific thought throughout our work. Science permeated our ways of thinking about our products. We used science to make estimates and predictions. If the predictions didn't work out, we designed and used additional scientific experiments to isolate the cause of the discrepancy – and it was always something we had failed to consider in our mathematical and mental models of the original design. We knew that the problems we faced had to be traceable back to some logically understandable phenomena.

We never failed. We loved this work. My friends and I produced excellent engineering results. Our customers were happy. My outlook was, and still is, that engineering success, and hence the success of a manufacturing business in general, is rooted in the inviolability of the laws of nature. The ability to invent a new product that has a good chance of working is based on being able to make an intelligible description of a new and untried combination of components, communicate the whole plan to a team and have each member responsible for a separate subsystem. The interfaces between subsystems are specified up front and fine-tuned as the work progresses. When the subsystems are all connected, and you turn on the switch for the first time, it just has to work. If it does not, of course, we KNOW it is the result of human error or defective components, and the problem-solving begins. We were practicing reductionists in this kind of activity.

Many of my fellow engineers – my teammates, my lifelong friends – were, and still are, deeply religious. Of those, some outspokenly describe themselves as "evangelical," "born again," or "charismatic." They speak freely of miracles, of prophecy, of revelation, of signs, of reasons, of the Holy Spirit. They pray for insight, to secure a blessing, to get spiritual help with a difficult personal decision, and to request divine intervention in Earthly affairs. They regard prayers, miracles, revelations and so forth as outside of science, but honestly believable.

They are certain that events on Earth can be started, stopped, shaped and controlled by higher powers not subject to scientific inspection or analysis, but simply viewed with a sense of awe. Many such people regard ancient writings as the very Word of God, and discuss them endlessly in an effort to improve their understanding and strengthen their beliefs. Their religions are extremely important to them, and are not treated casually. During the years we worked together on engineering projects, they held meetings in evenings or at lunchtime to discuss the Bible and other interesting writings.

I don't think they acquired this body of knowledge in engineering school. I doubted that they were using any of it to get their engineering work done, except possibly to obtain a level of encouragement and confirmation before trying a new approach to something. I often asked these friends how they could, with one part of the brain, rely completely and automatically on the known and tested rational laws of nature during the engineering part of their day, but then turn to the supernatural at other times, in effect asking for an unpredictable divine power to briefly modify those known and tested laws, at least in a small region of space. People had a lot to say, but nobody ever answered the question.

Is there a wall between one thoughtstream and the other? On one side of their brain, everything is rational, and follows known natural laws. On the other side, science is set aside, and its laws disregarded. Could these friends of mine, in fact, use their rational side to take a rational look at the supernatural sides of their own personalities? They could not. Though they could hop effortlessly from one side of the wall to the other, they could neither straddle it nor send a message from one side to the other.

There is, of course, a way to view supposedly-supernatural phenomena rationally. This is what our intellectual forbears have been doing for the past four or five centuries, and what led them to deliver to us today's laws of motion, gravitation, thermodynamics, optics, chemistry, fields and waves, quantum mechanics, genetics, evolution, *etc*. At first, a phenomenon is noticed, but not understood, for example the motions of the Sun and its planets. As was commonly done from deep history up through the Middle Ages, the phenomenon is ascribed by the common folk (and we are all common folk, regardless of our educational attainments) to divine providence, with heavenly beings driving the heavenly bodies around in their cycles. But some investigators eventually got curious, and got better equipment, and the questions begin to flow.

Creation of the Earth? Hmmm. It must have happened, but how? What temperature, what pressure, how long in the cooker? (For that matter, we will need to invent instruments that can help us measure pressure, temperature and time.) Was there a more complex protocol than simple heat and pressure? What's the recipe and the checklist? What materials were used? Where were they obtained? Were any of them dangerous to handle? Were there other preliminary experiments that led up to this? Have there been additional relevant experiments since this one? Did the original experimenter (assuming there WAS an experimenter) consider it a failure, a success, or just one step in a long, long process leading from a past unknown place to a future unknown place?

What were the experimenter's goals? How close did he/she come to achieving them?

All of these questions, and more, culminate in the ur-questions of science, "How does this really work in nature?" and more particularly, possibly blasphemously, "How could I reproduce that in my laboratory?" Work will continue as long as there are grants or wealthy dilettantes to sponsor the work.

Eventually the supposedly supernatural phenomenon acquires a logical and mathematical model consonant with other known natural laws. With the model in hand, people make trial predictions. If those never fail, we've got a usable law of nature in hand, a principle. We can make predictions with it that we could not previously make, and rely on them. "Never" is a long time, though, so we need to be always ready to reconcile the new principle with any new knowledge that presents itself. Science is self-healing. If there's a discrepancy, we can often offer a modification such as the one Einstein applied to Newton. Failing that, we can investigate a competing philosophical proposition and see if it accomplishes the necessary reconciliation.

I am at an early stage in my proposition (that the religious brain has separate compartments for religion and science with few neural pathways from each to the other.) I am confident that there is a rational explanation for this, and that it is consistent with other known principles in the biological and physical sciences. My prediction is that others who come after me will design controlled experiments, run them, publish, and expand our knowledge base.

So, those are some of the notions about how brains with and without the capacity for religion behave in observably different ways. I further assert that the stage is set for these behavioral differences by anatomical, structural, differences installed by genetics.

2 of 8: About 80% of the human population are religious, and about 20% are not.

OK, you may argue with the number, but you will agree that there is a number. I will show you the data I relied on. You may prefer other data, but the fact remains that the number of people who lack the capacity for religion, in all regions of the Earth, is nonzero. William James [James1910] referred to them as the "hosts of persons who cannot pray." It is difficult to measure what fraction of the world's population has the capacity for religious thought and action, but it is not difficult to determine that the number is substantial. How many of us have, and how many of us lack, the capacity for religion?

Measurements are unreliable. Surveys ask the question in various ways and get radically different numbers. Indeed, identical surveys that detect a few percent atheists one year may detect a significantly different number a few years later. People who lack the capacity for religion can be very cagey about whom they tell and what kind of question will tease the admission out of them. In many societies, the word "atheist" and its many synonyms invite prejudice, opprobrium, and physical hazard. In some countries, the legislative environments opposing freethinking are strong: in some countries, the death penalty is imposed on confessed nonbelievers of the state religion. Atheists know their place. Often, they may lie outright about their nonbelief, or hedge their replies in some way. The word "agnostic" was coined by T.H. Huxley (known as "Darwin's Bulldog") as just such a hedge in Victorian England. The counts reported by surveys, therefore, might be artificially depressed, but rarely inflated.

Survey results, as suggested above, are exquisitely sensitive to location, to political environment, and to the way the question is asked. A Wikipedia article current at the time of this writing [Wikipedia:Irreligion] allows us to compare responses to three different questions, country by country. A Gallup poll reports responses to "Is religion an important part of your daily life?" A Dentsu survey reports totals of those who replied that they have "no religion." A Zuckerman survey reports the number of responders who describe themselves as "atheist or agnostic." None of these, of course directly measures the presence or absence of the capacity for religion. Variation in the "no" response to the Gallup question ("Is religion an important part of your daily life") from country to country is exemplified by a few chosen points: lows - Indonesia 1%, Jordan 4%, Yemen 1%, Saudi Arabia 4%; Highs – Sweden 88%, Denmark 83%, China 82%; Mediums – Japan 71%, United States 36%, South Korea 52%. It is literally all over the map. It's notable that even in Iran, a famously strict theocracy, 8% found themselves capable of answering "no" to the Gallup question. Does that number truly represent the fraction of Iranians who lack the capacity for religion? Hard to say, since only 1% of Iranians replied "no religion" to the Dentsu question. Could it be that nearly all Iranians have the capacity for religion, but 8% of them don't make it part of their daily lives?

It would be useful to see the results of a survey question that could more closely reflect the presence or absence of an evolved capacity for religion in any individual; for example "Do you feel that your religion makes sense in the world?" or "Are you willing to fight to the death for your religion?" As regards the level of honesty we can expect in replies to these questions and any others, consider that there are many priests who are atheists, and many skeptical congregants who continue to congregate and go through the motions. A Web search for "priests who are atheists" returns about half a million hits. Social and family reasons make it very difficult to leave the fold.

These data points do not appear to directly support my contention that the prevalence of the capacity for religion is the same 80% everywhere on Earth. They do support my contention that both the capacity and the incapacity are distributed all around the Earth, though. A worldwide distribution supports the notion that the capacity arose in the hominin line long before hominins left Africa to begin radiating into other continents, and that it never achieved full 100% control of the genome anywhere on the planet.

My number of 20% comes from [Pew2012]. You might be persuaded by Greg Epstein [Epstein2004] that it's 15%, since the subtitle of his book mentions "a billion nonreligious people." Either way, there are people who will be surprised the numbers are so large. Visitors to the Pew report will see that the number of "nones" in the USA has risen from 15% to 20% in five years, and pause to reflect that such a large increase in such a short time is not likely to be coming from new immigrants or from a dying-off of believers. It could instead be coming from people newly discovering their inner lack of capacity for religion, or from people newly becoming more honest about disclosing what they previously knew but did not disclose. The Pew report adds another detail – that fully one third of all adults under 30 are religiously unaffiliated.

3 of 8: *Bona fide* conversion from either side to the other is rare.

Those without the capacity for religion, almost by definition, will never fully comprehend membership in a religion. Both "why" and "how" they should join a religion elude them. Any given nonbelieving individual may decide to make an outward show of going along with the flow for the sake of peace in the family or important social attachments, or to preserve life and limb; such values are important to all people, religious or not. The experience for the congenital nonbeliever, though, is not the same as that for a person who is born with the capacity.

For those who are born with the capacity for religion and are trained in a religion during childhood, there is similarly little chance of seriously moving away from the religion. Any given individual may step away from practicing the religion of his or her youth in order to make peace when marrying into a different faith or into an atheist family. These changes also find the person going along with the flow for the sake of familial attachments. They won't be fully satisfied in their own right, though, and should not generally be considered bona fide free will converts.

World history is full of stories of forced conversions. The most common scenes are those of religio-military conquest, and internal programs to convert "foreign immigrants" to an established state religion. Inquisition-style conversions may even be forced upon so-called foreigners whose progenitors had arrived in the territory many generations before the state religion did. Given a choice of death or acceptance – generally after a period of battle, incarceration, privation, torture or other indignity – people tend to outwardly accept the conversion. At least they say they do, since it is in their best interest to say so; they may have dependents who would be greatly inconvenienced by their deaths.

One wonders what life is like during the spread of a "new" religion, such as Christianity during its first several centuries, or likewise Islam in its multi-century wash across the Middle East and Mediterranean, and into Asia.

4 of 8: Religiosity and atheism tend to run in families.

If the capacity for religion is heritable, there should be some evidence of that. One clue is that a Web search for "hereditary atheism" produces five million hits. Individual stories abound of the form "I had a great grandfather and two uncles who were atheists." It's all soft evidence. It is difficult, unfortunately, to tease out the details at this time, for at least three reasons.

For one thing, people who are asked about their religious views do not all make honest responses. This is especially true of atheists (*i.e.* those who lack the capacity for religion) who fear harsh reaction from their community – and this fear is not unreasonable. It depends as much on the community as on the individual. It is also true of adherents to minority religions, for the same reason. I am certain that a great number of atheists simply report that they are religious, and go uncounted by the major surveys.

For another, sexual reproduction involves meiosis, and meiosis makes sure that no two gametes are alike. In this method of heredity, recessive genes for a minority trait can lie hidden for many generations, being passed on but not expressed, until the right sperm meets the right egg cell.

Thirdly, children receive training as well as genetic material from their parents. If a person reports that he is an atheist and so are his father and mother, is he reporting his training (which he might later depart from, if he actually possesses the capacity) or his heredity? Did he inherit the incapacity for religion from both of them, or was one of them enough? Was there a more subtle heredity path? Is the capacity for religion a dominant trait, or recessive? How many genes are involved in its transmission?

Web searches turn up many stories and dialogues about the heritability of atheism, which to my way of thinking is the other side of the same coin as the heritability of the capacity for religion. Unfortunately, they offer little evidence either for or against the notion that the capacity is hereditary. No other method of transmission has suggested itself, though.

5 of 8: The capacity established itself before the African diaspora.

Let's explore the notion that the capacity for religion was established in the human genome before human stock began streaming out of Africa to go populate the Earth. This process is variously estimated as having begun 100,000 to 250,000 years ago. The capacity for religion is present in all races aboriginal to all continents, and so is the incapacity. See [Frazer1922] and {Chagnon1968} for examples of its expressions in ancient forest cultures. It is present in all such cultures, even those with radically different characteristics. It is present in all lines of descent, even those known to have been isolated from other lines for many tens of thousands of years. If the original emigrants from Africa didn't have it, it would have needed to evolve separately in separate lines of descent on separate continents. It is indeed possible that this happened, *i.e.* that it evolved multiple times in multiple places. The principle of Occam's razor – that the simplest adequate explanation is preferred to more complex ones – suggests that the theory of a single-point origin with worldwide radiation is more credible. If the capacity actually originated in all continents separately, wouldn't it be even a tiny bit likely that one continent might have been overlooked? But they all have it. See also [Frazer1922] for an early worldwide survey.

When we get to the mechanism of the capacity's primal expansion, in Chapter 4, we will explore this question again and hopefully put it to rest.

6 of 8: Most People Grasp the Notion of The "God-Shaped Hole"

Most people, atheist and believer alike, readily understand the metaphor of a "god shaped hole" in the mind. It represents the supposedly universal human need for personal contact with gods and other supernatural entities. A Web search for "god shaped hole" produces over 2 million hits. Atheists, of course, will assert that it is not universal at all – and, in fact, it doesn't exist for them. But at least they know what the phrase means. They know that the "hole" exists for some people, and that any person they chance to meet may have it. Some atheists regret their inability to access this element of human sensation, knowing that others visit it frequently and that it brings them great joy. Others, despite not having experienced it, believe it would be seriously upsetting to them, and are just as glad that they are barred from it.

Some writers, like the aforementioned 17th century polymath Pascal, called it a "god-shaped vacuum" and placed it in the "heart." Today's thinkers call it a "god-shaped hole," as we just did, and place it in the "brain." I tend to side with the latter. It is a universally recognizable notion for what I have been calling the capacity for religion. People with god-shaped holes in their brains and people without them will take completely different approaches to the thoughts and actions of religion.

7 of 8: We have in hand a workable evolutionary explanation

We have in our hand (*i.e.* in this very book) a plausible explanation as to how a dedicated instinct – a capacity - for religion could have been reinforced in the human brain. The reinforcement by mortal combat and two other capacities will be expanded considerably in chapters 5 and 6. Chapter 5 will include an equally compelling explanation for why there are atheists remaining in the population, despite the great differentially-adaptive power of the combined capacities for religion and mortal combat.

It would be irresponsible for anyone to dismiss these interesting explanations simply for their speculative novelty, or for their lack of peer-reviewed concreteness. Time and hard work will be needed to fill the gaps, shore up the tunnels, and repair the defects. In the end, we may find that this explanation fails, despite all the time and hard work we put into it, but we will then know exactly HOW it fails. The work is useful either way.

8 of 8: We abjectly lack a competing explanation

Given that I've done what I can to describe these aspects of the human situation with reasonable vigor and completeness, can you come up with a more workable explanation for how it came to be? I will be gratified if all I have provided here is a straw man for someone else to knock down with a more powerful story. We shall hash out our differences amicably.

In the next chapter, we will explore the evolutionary economy of religion. I will suggest that the capacity for religion should not have evolved on its own, but it did. Did evolution have a breakdown?

> "natural selection is ruthlessly economical: traits, particularly energetically costly ones, do not evolve unless they serve some function." – Dorothy Cheney

4. Evolution Fails To Remove the Capacity for Religion

Evolution by natural selection, to borrow four adjectives from the Scout Law, is trustworthy, thrifty, brave and clean. What concerns us in this chapter is its thrift: evolution either removes heritable features that cause the organism to incur unnecessary costs, or modifies them until the costs become tolerable. It can be trusted to do this bravely and cleanly. This chapter argues that the religions that ride on the capacity are costly, perhaps so costly that the capacity could not have evolved on its own. It did evolve, though, so putatively not "on its own." Evolution produced it, just as it produces every other characteristic of every animal or plant, and has so far failed to remove it.

How Expensive Is Religion?

Is religion expensive? Short answer: yes. How expensive, and in what currency? Long answer: consider evolution's necessary preference for numerous, well-nourished and well-educated offspring. Imagine two nearly identical species inhabiting the same general countryside and enduring the same shortages year upon year. One is tuned, by a system of innate capacities, to maximize its delivery of resources to its offspring. The other is tuned, by a system of innate capacities, to devote more resources to issues of interest only to adults. But all resources in this countryside are equally scarce, and there are no surpluses. Choosing how to spend resources is the proverbial zero-sum game. Which of these species increases more rapidly than the other?

Of course, the situation is far more complicated than that, but the basics hold. A million or so years ago, on the savannahs of Africa, there was a silent budget – a budget of scarcity – constantly at work on one pre-human species after another. Did it reward an investment in peaceful growth, or an investment in beautiful rituals, grave goods, monastics and priests? We will discuss a different algorithm in the next chapter. For now, let's enumerate some of the resources that religion consumes, resources that evolution would prefer to see invested in population growth.

A typical religion meme, hosted by the heritable neural capacity to be religious, diverts resources to itself, away from physical benefit to the population that would increase its growth. For the sake of our religions, we take land out of agricultural and commercial service. We build enormous structures at great expense: in support of the point involving large structures, consider that of the original "seven wonders of the ancient world," (See [Cottrell1962]) only one, the Pharos of Alexandria, directly benefits human welfare. It aids navigation and prevents shipwrecks. The other six are flights of fancy. One (the Hanging Gardens of Babylon) is a monument to love – it possibly benefitted human welfare by increasing the builder's physical intimacy with his morganatic mate. But the other five – two tombs (Khufu's Pyramid and the Tomb of Mausolus), two statues of gods (The Colossus of Rhodes and the Olympian Zeus) and one vast temple (The Ephesian Artemis) – are totally religious in their plan and purpose, and provide no service to the biological economy of the species.

We feed, clothe and house myriad men and women who spend their time in religious contemplation and philosophizing rather than in bearing children and teaching them how to cope with the physical world more effectively. We bury valuable and useful objects alongside dead bodies in tombs as grave goods, never to be used, exchanged, or even seen again until discovered by archaeologists. We house, clothe, and nourish non-reproducing mendicant pilgrims on their lonely marches hither and yon. We experiment with monastic and priestly celibacy, which removes large groups of men and (especially) women from the cutting edge of the need to breed. Our secular governments allow religious businesses to operate and accumulate wealth without taxing their gains. Adherents will their life savings, properties and other accumulated resources to their religious establishments, rather than to the education and biological recruitment of their young. We tithe. At our altars and temples, we make ceremonial sacrifices of foodstuffs and other valuable resources. We maintain standing military forces that consume resources while idle, and then risk and sacrifice their very lives in order to defend our faith or to attack other armies that defend different faiths. We ask permission of religious authorities before publishing our thoughts on natural philosophy (see earlier remarks on NoMa). Some religious leaders overtly retard intellectual progress in science. Still others interfere with the administration of vaccines and other life-saving medicines to children. We surgically modify our children's genitals. We impose dietary restrictions on religious adherents, and regulate food preparation methods in ways that increase cost, but provide no nutritional enhancement. We write, edit, copy, publish, and store sacred writings and theological essays. We maintain a huge market in buying and selling those writings. At life's passages, we stop to conduct elaborate rituals in order to secure the approval of our supernatural community.

The notion of evolutionary expense does not originate with me. A Web search for "costs of religion" produces 98 million hits, far too many for anyone to follow. Somewhere in that pile is an analysis that gives real numbers.

In presenting to others that notion of religion having costs that should have kept it from evolving, I am often given pseudo-evolutionary rebuttals, such as

Religion evolved in order to keep us from killing each other;

Religion evolved because it offers compulsory gatherings, and so brings us together and helps us find marriage partners;

Religion evolved because we need morals, and it is the source of all our morals, *e.g.* the Ten Commandments;

Religion evolved because it provides an education structure, so that more children can be more effectively trained in the skills and cultural tenets they need to succeed.

I am grateful to my friends for these suggestions. Some were given in a paradoxical context, *i.e.* offered by young-Earth creationist individuals who also reject the natural law of natural selection to begin with. But I don't wish to base any of my assertions on *ad hominem* arguments. Instead, let's examine the suggestions one by one, and see if any of them, or any combination of them, is enough to repay the costs discussed in the 98 million Web hits described earlier.

First, we seem to kill each other a lot, even with religion in our blood and bones, don't we? Would we kill each other even more if we didn't have a capacity for religion? Oddly enough, exactly the opposite may be true – religion may actually have been helping us to kill each other throughout history and prehistory. Hold that thought: we will have a LOT more to say about it in chapter 5.

Second, human societies have many nonreligious ways of bringing groups together in ways both structured and unstructured. Examples are marketplaces, political meetings, music and drama performances, group dances, athletic contests, pancake breakfasts, schoolrooms, neighborhoods, military recruitments, parades, leks and the like. Our very rich and evolutionarily ancient coalitional capacity could pretty much take care of getting us to form up in groups, even when not assisted by our religion capacity. I'm sure it once did, and still does. I will describe in chapter 5 that our coalitional capacity is vastly older than our religion capacity. The argument will be based on chronological phylogeny of monkeys and apes – these animals, with which we share common ancestors at reasonably well-estimated ages, are coalitional, but not religious. Their methods of finding mates seem to be perfectly adequate. Does adding a supernatural layer to a compulsory gathering increase the gathering's effectiveness as a generator and nurturer of offspring? I think not. It seems to me that we can – and do – meet just as effectively for mate-finding and for satisfying other social needs without adding myths about the adventures of supernatural creatures and entrusting their telling to the priesthoods, their appointed human representatives here on Earth.

Third, the strength of the connection between religious education and the capacity for moral thought and action has long been widely overstated. It's doubtful that the onset of moral thought and action in a maturing human requires any precipitation by religious education. It's very likely, though, that the capacity for morality arose along with the capacity for coalitionality. A coalition lacking an inner sense of morality would be a dangerous living group in any case, presence or absence of a capacity for religion regardless. Both capacities, as will be brought out in chapter 5, are much, much older than the capacity for religion. Our distant predecessors the monkeys had morality without religion, and so do 20% of today's humans. To look further into monkey morality, search the Web for "monkey morality" and get half a million hits. To look further into atheist human morality, search the Web for "atheist morality" and get nearly a million. Some useful references relating to evolved morality are in the Bibliography: see [Epstein2004], [Katz2000] and [Wright1994].

Our social confusion on this point is due to persistent efforts on the part of priesthoods to justify their livelihoods and preserve their authority. All well and good, but there's little evidence to show that religious training makes anyone more moral than secular training does. Indeed, individuals are known who have religion but lack morals, and *vice versa*. Golden-Rule style moral codes appear in and out of all cultures over all time, which fact suggests that the moral codes arose long before any of today's extant religions. Each new religion claims fresh authorship of the old moral codes, in effect reinventing what was already known and practiced by previous religions and by nonreligionists as well.

 Although much is made of the teaching authority of the Ten Commandments, close inspection shows that half of those dicta are neither observable nor enforceable, being purely mental in their scope. It is impossible to know whether or not another person is having impure thoughts unless he or she tells you. People generally know which of their thoughts to share and which they'd rather keep to themselves. All of us know which of our thoughts are pure and which are impure: what evolutionarily sensible reason would there be to reveal one's secret impure thoughts? The other Commandments – the ones that actually proscribe evil deeds that bring harm to others, are all directly derivable from the Golden Rule: their moral tenets, being innate, are obvious and agreeable to nonbeliever and believer alike. To probe further in to the innate instinct of the Golden Rule, search the Web for "inborn golden rule" and get 24 million hits. Is any moral enforcement power gained by having priests teach us that moral rules were once written in stone and handed to a human by a supernatural being? No. They were written in DNA, not in stone, and by evolution, not by a supernatural being. Evolution does not care how the rules are taught, because those rules are innate and do not need to be taught. If any teaching were needed, what would be the benefit of calling on a supernatural authority while doing the teaching?

In any case, back to the question of the evolutionary advantage of the capacity for religion. If some activity we observe has no effect on the improvement of reproduction, the evolution algorithm will neither increase it nor reduce it. How would the teaching of the Ten Commandments bring about increased reproduction among the taught, relative to that of the untaught?

Fourth, about schooling: secular and religious schools have coexisted for millennia. To the best we know they are equally effective at initiating the young into the important parts of the cultures that surround them. People who come through either educational path are similarly equipped to deal with social life. The supernatural aspects of religious education are of personal value only to those with the capacity for religion. Nobody has yet suggested to me the mode of action, *i.e.* how does religious education increase fecundity?

What Pays Religion's Costs?

In preparation for Chapter 5, consider this: wherever evolution produces something complicated or expensive, or just makes you scratch your head, there's bound to be something else at work that helps it out. We will discuss the "something else" – coevolution with the capacity for mortal combat – in chapter 5.

> "... the very notion that one can develop any new explanation reflects the confidence that one can tackle a problem whose solution has eluded so many others before. ... I will count myself successful if I have prompted scholars to think in new ways about religious violence." – Hector Avalos

5. Coevolution with the Capacity for Mortal Combat

Religion Makes a Man a Better Fighter.

Human brains, as noted throughout this argument, are products of Darwinian evolution. This indisputably includes yours and mine, and those of every believer and every nonbeliever as well. It includes those of every past, present, and future human child.

Mortal combat – man against man – is a recurrent situation in our evolutionary life. It is very ancient. It occurs worldwide. It is faced by every generation in every country. I will assert that in the human genome there is a heritable capacity for mortal combat, and that it is older than the capacity for religion.

Each individual incident of mortal combat on the savannah of ancient Africa – man against man – lasted but a moment: yet its outcome was irreversible, and the evolutionary result has lasted to the present day. The cumulative effect of combat after combat, like compound interest, repeated and recursively multiplied over millions of years by billions of killings of billions of men by other men, makes it a crucially important element in our evolutionary history. It also gives us pause to consider the fact that we are still doing it, despite what would seem to be adequate reason to stop doing it.

As noted in chapter 2, there is fossil evidence of mortal combat. Let's examine what it might mean. There are approximately twelve extinct species of hominin known to science from their skulls, bones, tools, grave goods and the like. The number twelve is a matter of controversy due to the scanty nature of the fossil record and the close resemblance of each species to another. Experts argue with other experts whether fossil A is the same species as fossil B, and whether or not fossil E is in the same line of descent as fossil C and so forth. See

[Bryson2003] for a longer explanation. We are pretty clear on the ages of most fossils we find, as many of them are found buried in sedimentary strata or volcanic ash deposits that can be accurately dated. The familial ties, though, are not so clear. Fragments of nucleic acids may yield some clues. It's probable that the line of our descent is not a simple connecting of the dots from the oldest known hominin fossil down through the more recent ones to us. Some finds may represent side branches, evolving at a time when the world tolerated more than one concurrent species of hominin. But the sparsity of the hominin fossil trove is one of the things that makes it possible for us to state with some certainty that mortal combat has been with us a long time. Though we have only a few fossils to study, many of them exhibit combat damage.

Think of it – twelve known (and how many more that are not yet known?) extinct antecedent species of Man going back a million years, a large percentage showing evidence of combat injuries, and only one species of Man extant today. We are (to borrow a pliable catchphrase) the "last Man standing." What were the circumstances of the twelve extinctions? What will be the circumstances of our own extinction? This does not sound like a difficult puzzle. Although some of you will say "no way could that have happened," it appears possible – even likely – that each predecessor was annihilated in mortal combat by a somewhat improved successor species.

Mortal combat is a guiding force in the development of our genome. Our species survives it and keeps it handy. Our species survives and develops in our niche – despite combat, some say. It's also possible that we became what we are – essentially nicheless – because of it. We've conquered the entire globe as our niche, and are making moves on the Moon and Mars. The twin facts that we have survived despite having a capacity for mortal combat, and that we consider it normal, seem to indicate that we are comfortable with our capacity for mortal combat, and evolutionarily adapted to make the very best use of it that we can.

Adaptations like the capacity for mortal combat are not quantum-mechanical errors, though they may result from such accidents at their inception. Their continuing development unquestionably results from the inexorable algorithmic process of natural selection. How, in detail, does this work? In that brief moment where defeat and victory meet, a life beats a death. One man's potential genetic contribution to the ongoing and developing gene pool is abruptly extinguished, and another man's is preserved – at least until his next combat incident.

The human is the only mammal species (indeed the only animal species) known, in which individual members regularly and intensively engage fellow members of their own species in mortal combat. Evolutionists who look at evolution in general as a continuous gradual process driven by environmental stresses, predation, parasitism and asteroid impacts are apt to overlook the importance of intra-species mortal combat. There are three reasons I can think of for the oversight: first, because we spend so little of our time actually engaged in it; second, because of its rarity in the Animal Kingdom – only one other species does it at all, and to nowhere near the degree that we do it. A third reason is that when we are not actively doing it, we are embarrassed by it. We consider it ugly, and would rather not have it as part of our makeup. We are much more comfortable thinking about extinction by falling asteroids than we are thinking about the capacity for mortal combat evolutionarily shaping and refining large areas of our personalities.

Mortal combat as a driver of evolution is not the usual, familiar, textbook kind of "nice" gene shaper. This evolution-driver isn't the result of uncontrollable external environmental stresses. It's the result of internal impulses and processes that originate within our own brains. It is fast and effective. It results in the immediate settling of a dispute in the gene pool.

A combat killing is fundamentally different from a killing by a predator: if a man is caught and eaten by a crocodile, his genetic contribution is terminated just as it would have been by his death in mortal combat; but the crocodile's genes don't enter the human gene pool. Likewise, if the man is lucky enough to defeat the crocodile, the man's genes don't enter the crocodilian gene pool.

We must no longer ignore the workings of mortal combat on our gene pool. Combat has had an immense impact on the formation of our entire species. Its binding with the capacity for religion occurred in ancient times, and it is now no longer an evolution driver. Nonetheless, the combination is still present and still highly active.

The Mutual Binding of Separate Capacities

The evolutionary impact of intraspecific mortal combat is fast and big, *i.e.* killers beget more killers, nonkillers beget fewer nonkillers. This action shifts the gene pool's bias toward more killers and fewer nonkillers. It tends not to let the balance ever drift back toward nonkillers – how could it do that? What would the nonkillers' adaptive advantage be? But there's more to it than that. The capacity for combat, so quick and so strong, won't be left alone. Other capacities that aid and abet it will inevitably hitch a ride on it, and it on them. Some might say they "gang up." As the capacity for mortal combat amplifies itself, it also amplifies any other traits that have become attached to it. The genetic contribution that leads to mortal combat recruits into partnership the genetic contributions for all other heritable capacities that have a bearing on the outcome of a fight to the death. The capacity for mortal combat will magnify or diminish such traits, along with itself, according to their utility in winning events of mortal combat. A trait that helps a man win a lethal fight will be amplified in successive generations along with the capacity for mortal combat itself. A trait that interferes with victory in combat will be correspondingly diminished.

Any heritable trait that adds to a man's ability to succeed in mortal combat cannot escape being magnified in succeeding generations: by the same token, any trait that reduces success in mortal combat will be diminished over generations.

With all that as background, consider together the capacities for religion, *i.e.* the heritable mental apparatus that enables religion in the human brain, and the capacity for mortal combat. The capacity for religion became entangled with the already-existing capacity for mortal combat in deep prehistory, just as soon as it appeared in the genome. Each aided and abetted, and fed mutually, on the other.

How does the capacity for religion act to strengthen a fighter? I've postulated earlier that the capacity for religion makes it possible for a man to think and act in ways that are not fully rational. Under some circumstances, life itself, it seems, can disconnect from rationality, and proceed on an emotional autopilot. Mortal combat is such a circumstance. I postulate here that, in the instant of mortal combat, the man whose mental processes are fully rational is at a distinct disadvantage.

The rational man's opponent, equipped with the twin capacities for mortal combat and religion, has powers in battle that the rational man lacks. To name a few,

- he's fighting for the glory of a higher power;
- he is filled with the personal power and glory of that higher entity;
- he's defending the faith against the infidel, that godless, subhuman animal;
- the lord will frown on him if he fails to give 110%;

- he's guaranteed an excellent, immediate, permanent afterlife if he falls in combat;
- he'll get a hero's reception back home if he prevails;
- the hero's reception includes abundant sexual favors;
- those favors provide increased reproduction of his inherited capacities;
- he's guaranteed a life of shame and non-reproductive isolation if he fights in a cowardly fashion;
- he is here to fight and kill the enemy, not to think or negotiate with him;
- "not to reason why, but to do and die" (Tennyson, *The Charge of the Light Brigade*)

That man is a ball of emotion all aimed at fighting to the death. Life (his) is on the line. It's a must-win situation.

His opponent, the fully rational man, lacks one or both of the twin heritable capacities for combat and religion. He has been thrown into an irrational situation, and is therefore off balance to begin with. He:

- does not fully grasp what's important about the immediate present;

- may be slow to realize that he really, truly must take the life of another man in order to save his own;

- may not have previously faced the instantaneous need to preserve his own life (a tautology, because he probably would not have survived such a previous engagement);

- may waste time gathering data, building a list of alternate approaches to the problem at hand, comparing and choosing among them (We have been taught that rational thinkers do this before proceeding to propose a solution, haven't we?);

- may fruitlessly propose his choice of solution to his enemy;

- expects he will be overwhelmed with guilt if he kills a man.

This is not to say the religious fighter always wins, just that he has faster, more decisive fighting equipment in his brain. All other things being equal – and they rarely are, but they average out - he is more likely to kill than to be killed.

Recall also, from the discussion of the Trolley Problem, that instinctive processes tend to be faster than stepwise reductionist practices. This principle further accelerates the sword of the religious fighter. If you were a gambler, you would do well to bet more on that man than on his imperfectly-equipped opponent. And, in fact, most of us do, unwittingly, make that bet. We bet the future of our species on the rightness of our genotype each time we conceive a child. The only way to opt out of that bet is to opt out of reproduction, and leave the arbitration of gene pools to others. That's still a bet, though, isn't it?

Coevolution with the Capacity for Mortal Combat

In the chapter 4, we discussed the cost of having to support a religion meme. In this chapter, we are describing the notion that the capacity for religion has a helper. It has its expenses paid by a supporting companion capacity, that for mortal combat. The costs incurred by the human being having to support a religion meme are more than offset by the advantage that the religious mind confers on the warrior engaged in mortal combat.

The fact that the advantage is differential rather than absolute must also be discussed. If the deleterious capacities that we use to kill each other off don't outpace our birthrate, the species as a whole advances to the next square on the game board. If the kill rate outruns the birthrate, population by definition decreases, but chances are good that it will recover. Once the age group that normally engages in mortal combat is depleted, combat tapers off until the juveniles grow to recruitment, that is, to fighting age.

Consider the universality of mortal combat yet again. It's a constant preoccupation of our species, right after food, clothing and shelter. I'm sure many readers have actually participated in mortal combat. Every one of us has known someone, or is related to someone who has participated in mortal combat. Many of us have lost loved ones to mortal combat. We erect great monuments bearing the names of our dead countrymen who fell in mortal combat. Men of all nations in all generations have been invited, and even compelled, to join in mortal combat. We read about mortal combat in the daily papers and see it on the nightly TV news. In these modern times, nations invest significant portions of their resources in "defense." We are always ready.

In chapter 4, I noted the paradox of the apparent nonadaptiveness of the capacity for religion. The capacity for mortal combat is not so paradoxical. The actual performance of mortal combat, like the performance of religion or language, is a meme supported by an underlying neural capacity. Without the capacity, mortal combat would be nonexistent, or at most, fitful and temporary, and always defensive. Military training would be ineffective. Invasions would be impossible. Like the capacities for religion and for language, the capacity for constant mortal combat is unique to our species. Killing has been observed in chimpanzees, which are not of our genus or species, but are of our hominid family, the great apes. The common ancestor of chimpanzees and humans is thought to have lived between six and nine million years ago (we have not yet found a fossil of that line, so rely on calibrated rates of change in mitochondrial DNA).

Chimps have been observed to practice mortal combat occasionally, incidentally. We, the so-called *Homo sapiens*, practice it regularly, universally. We have it in our automatic background. When we are not personally involved in a particular episode of it, we tend to stop noticing that it's going on all around us. At the time of this writing, there are over 40 shooting wars in progress on five continents of our planet, causing over 100,000 violent deaths per year. The number changes frequently, of course, so repeat the research often to get the latest. My source is Wikipedia, but there are many more.

Since this chapter is about the marriage of two instinctive capacities – the capacity for religion and the capacity for mortal combat – I shall now spill the rest of beans. There are at least two more partner capacities that operate in a cycle of mutual enhancement. Chapter 6 introduces the human capacities for genocide and kleptogamy.

Prepping For Battle

Let's examine the activities that precede and follow episodes of combat. The men of today do things that are in some ways similar to things that were done half a million years ago in the Savannahs and forests of Africa. We have some of the same instincts. These are evolutionary echoes, bubbling up from our legacy library of ancient evolved capacities. In preparing for combat, men of each band (coalition) assemble. They decorate their bodies with special clothing, paint, ribbons, talismans, phylacteries and armor. Fierce traditional insignia are adopted, images that commemorate past victories or losses to be avenged. The fighters-to-be conduct ritual exercises, sing songs and recite heroic stories, myths and legends that have the effect of binding individual coalition members into a coherent, dedicated force. These same stories begin the process of demonizing, demeaning, denigrating and besmirching the enemy whom they will soon engage in combat. The enemy is reduced to a less-than-human status, a mere lower animal, an insect, a disease; he is thus rendered ineligible for clemency under the universal Golden Rule. The fighters recognize and accept their leaders, and generally accept their positions in hierarchies of leadership, if any. They train, exercise and strengthen their bodies and brains to meet the challenges to come. They may form specialist groups. And through it all, if they have the capacity for religion, they pray, worship, glorify and propitiate their special supernatural deities, asking for success in the risky fighting, promising ever greater glorification after the great victory to come. They pooh-pooh and otherwise minimize the powers of the enemy's false and superficial religions. The opposing army's deities themselves don't really exist, and even if they do, they can't effectively compete with our team's. The enemy's totem animals are worms and insects. The things the stupid enemies say they believe in are nothing but paper gods, phonies, trivial inventions of the cynical bosses of the sniveling unclean enemy people/animals. They aren't real gods, not like ours are. The enemy's faith in their gods is itself weak. The faith, if they have any, is false and erroneous. Their falseness, error and hypocrisy will bring upon them the wrath of our true, real god(s), who will facilitate the enemy's destruction. All we have to do is swing our axes in His service.

On command, they go forth to the stirring sounds of bagpipes, trombones and drums, to begin the battle. In each hand-to-hand combat, may the best man win. Literally, may the fittest survive.

It is easy to imagine how a capacity for mortal combat might have spread rapidly through a whole human population on its own. As the males of the first combative band engaged males of a noncombative band, the outcome was not in doubt. It was the immediate end-time for the noncombative genome, and party time for the combative genome.

Immediate Aftermath

The consequence of such a biased victory is a profound one. Dead men tell no tales, it's said. They also produce no offspring. Offspring will be sired only by the living men, *i.e.* by the victors. History is written by the victors, it's said, and the important part of that is written into the post-combat genome of the next generation of the species. Religious men are the more likely victors, having the advantages of an inner fire, a higher cause, and a capacity for automatic, instinctive, automatic killing action that requires no thought and consumes no time.

Battle heroes are more than mere men: they are fountains of DNA. Being victors, on returning to their home village they are venerated by their own band, which may be slightly depleted of men itself. The veneration, in turn, leads the heroes on to venery. That in turn results in the gametes of the more heroic fighters being delivered to a much larger cross-section of the band's fertile women. The magnification of one hero's nucleic acids into succeeding generations can be immense. A living example has been plausibly identified, involving male descendants of Temujin "Genghis Khan" (see [Mayell2003]). The great Mongol chief's personal Y chromosome has been identified in 8% of the male population in the region he once ruled, or ½% of the entire male population of the Earth. It has only been eight centuries since his time in the world. This underscores my suggestion that combat-driven changes can sweep rapidly through a region. If you don't agree, you are obligated to supply your own ideas as to how this might have happened.

Chapter 6 will be devoted to the fate of the civilians, women, and children in the losing band, but for now mainly note that each battle, while reducing the total number of males, has not changed the total number of females: no females from either band were engaged in the combat. Their lives were not risked. The reduction in male population from all causes including mortal combat is nominally immaterial to evolution's scorekeeping, because it is the size of the female population that governs the band's overall potential reproduction rate. A large population of females can be kept pregnant by a very small number of men. See [Mayell2003] for a description of what is known about the immediate offspring of Genghis Khan and their harems and concubines, with implications for a staggeringly high ratio of females to males engaging in procreation. The penetration of the capacity for religion is thereby further amplified. One way or another, the next set of offspring will be preferentially sired not by the original (putatively nonreligious) members of the losing band but by the putatively more religious men of the victorious band. In Chapter 6, we will see that the capacity for kleptogamy ensures that the losing band's women will not be excluded from reproductive opportunities.

The Splitting of Consciousness

In earlier chapters, more than once, in fact, I mentioned that I had a long career working with engineering teams in which some of my teammates clearly possessed the capacity for religion. They also had the capacity to set their religions aside while doing work that required full rationality. I assert that this tendency to shift easily back and forth between scientific thinking and magical thinking is universal among people who have the capacity for religion. The capacity for religion is, among other things, a capacity to develop two sides to one's consciousness, and use the appropriate side at the appropriate time. People who lack the capacity are stuck on one side forever, and are forced to think rationally (as best they can) all the time, even when it's not in their best interest to do so.

They live acceptable lives, though bereft of the pleasures of supernatural magic, and likely unable to succeed in mortal combat. It seems to be possible to get along in the world without having the capacity for religion, and without having to fight for one's life, if one avoids certain situations. In fact, all animal species, save one – ours – successfully occupy their niches without religion and without mortal combat. But it would be hard to imagine an animal getting along in the world on pure emotion and magical thinking. Even to those with the capacity for religion, some degree of cause-effect reasoning must always be available for immediate use at appropriate times. Rationality cannot be permanently erased from one's life, even though we know that there are moments such as mortal combat where rationality is counterproductive. Imagine daily life totally without rationality. I'd say it's unlikely that the capacity for religion would have met with evolutionary success, if it had caused the brain to completely discard the rational processes with which it shares the living brain. The successful capacity for religion must allow rational thinking to resume once the combat ends and the threat tapers off.

The splitting of consciousness, then, is an important facet in the shape of the capacity for religion. Though the mental machinery of religion holds its supernatural truths to be universal, permanent, absolute and triumphant, it is willing to move them into a background role whenever it can get better results from the body of natural laws that continue to emerge from the processes of science.

Summary of Chapter 5

The capacity for religion, including its tendency to sagely balance one set of thinking against another, hitched a ride with the capacity for mortal combat in ancient times. They are two of the simple rules that create the complex result (ref. Chapter 2) that we call *Homo sapiens*. As a joined pair, they have moved into the fast lane on the evolutionary highway. One's religious side might say "the god of victory has prevailed:" one's scientific side might say that "the algorithm of natural selection made it so." The two sides would not be in disagreement. After all, the two capacities live in one brain, and share all the physical body systems of one individual.

"But time and time again, decent men and women chose to look away. We have all been bystanders to genocide." - Samantha Power

"The headman of the group organized a raiding party to abduct women from a distant group. They went there and told these people that they had machetes and cooking pots from the foreigners who prayed to a spirit that gave such items in answer to the prayers. They then volunteered to teach these people how to pray. When the men knelt down and bowed their heads, the raiders attacked them with their machetes and killed them. They captured the women and fled."
– Napoleon Chagnon

6. Additional Accelerants

By this point in the book, I have pulled out all the stops to show that the capacity for religion and the capacity for mortal combat are entangled as a coevolutionary pair. The capacity for mortal combat almost certainly predates the capacity for religion, since the capacity for combat would still prosper in the absence of the capacity for religion. Religion, on the other hand, would be unlikely in a species without mortal combat. What would it do by itself? This pair of capacities rides on top of the far more ancient capacity for forming coalitions, so we have attained a view of a coevolutionary trio by this point.

In this chapter I introduce two more capacities that coevolve with the trio to form a quintet. The first of the additional accelerants is the capacity to commit genocide, especially the capacity to massacre or neuter and enslave males of an opposing coalition. The second accelerant is the capacity for kleptogamy – the stealing of women from an opposing coalition. We shall examine examples of such occurrences ancient and modern, show that they are very common, and explore ways in which evolution would favor the procreation of individuals whose genotypes included these two additional, and most regrettable, capacities.

It is at this point in the book that you will see the reason why I advised you to hold the work at arm's length, regard the subject as some abstract animal species, and ignore for the moment that you are a member of that animal species. Despite its innate Golden Rule, our species is capable of great and savage barbarity. Enough said – now we continue as before.

The scenario for this chapter, like that for the previous chapters, is our deep, ancient time on the vast African savannah. Humans have been busily descending from their predecessors for a few million years, forming small bands of semi-nomadic hunter-gatherers. Bands, or emissary groups from bands, go forth to meet other bands to conduct trade, forge alliances, and carry out raids. As surmised in Chapter 5, each extinct version of genus *Homo* met its end by getting annihilated in combat with a later model carrying a new capacity or physical modification. Maybe the roving-nomad model does not create enough violence to bring about an annihilation of another band, though. Could there be other capacities at work here? Combat only kills those who directly engage in combat. What if there's another capacity at work that results in the deaths of noncombatants as well? We've discussed how the capacity for religion intensifies combat, and thereby reduces the population of those who don't fight particularly well. But we have not yet examined how religion-fired combat, aided by a capacity for genocide, might result in the mass annihilation of those on the losing side who didn't even fight at all. Imagine that being a regular event, and read on.

The Capacity for Genocide

We are all familiar with the term, which was introduced to the English language in 1943 by Raphael Lemkin *q.v.* (Wikipedia) as "the destruction of a nation or an ethnic group." The term was adopted by the United Nations in 1948 and ratified in 1951, after the addition of much fine structure, including:

(a) Killing members of the group;

(b) Causing serious bodily or mental harm to members of the group;

(c) Deliberately inflicting on the group conditions of life calculated to bring about its physical destruction in whole or in part;

(d) Imposing measures intended to prevent births within the group;

(e) Forcibly transferring children of the group to another group.

Broader aspects of genocide which interest us here include, in Lemkin's words,

> "Genocide has two phases: one, destruction of the national pattern of the oppressed group; the other, the imposition of the national pattern of the oppressor. This imposition, in turn, may be made upon the oppressed population which is

> allowed to remain, or upon the territory alone, after removal of the population and the colonization by the oppressor's own nationals."

Lemkin, asked why his interest in the subject was so sustained and so intense, replied:

> "I became interested in genocide because it happened so many times. It happened to the Armenians, and after the Armenians, Hitler took action"

A Web search for "genocide 2012" gave me nearly 12 million hits. Near the top are contributions from Genocide Watch [Genocidewatch2012] including an annually updated document that describes genocides in various stages of execution, country by country around the world. Lest anyone be under the misapprehension that genocide is a rare horror occurring once or twice a century under extreme and infrequent conditions, have a look at the Genocide Watch documents. The 2012 document cites 48 regions of 36 countries that are in stages 5, 6, and 7 of the 8-stage sequence of events that comprise a genocide. The 8-stage model, an elaboration of Lemkin's two-phase model, is composed of:

1. Classification
2. Symbolization

3. Dehumanization
4. Organization
5. Polarization
6. Preparation
7. Extermination
8. Denial

For complete paragraphs that describe each of the 8 stages in detail, in some cases along with suggested preventive measures, see the PDF associated with the Genocide Watch Web page. Stage 8 ("Denial") is noted as being not a completion of the cycle, but as a sign of genocide's deep, deep roots. It is said to be "among the surest indicators of further genocidal massacres."

But that's just 2012. Perhaps 2012 was a particularly genocidal year? No. Do a Web search on "genocide 20th century" and receive 2 million hits. One, [Levinger2006] reports that over 170 million people were "murdered by governments" during the 20th century, and those are just the ones conducted by "governments." One is led immediately to wonder how many more were murdered by "insurgents," "militias," or "tribal rebels," or some other such euphemism for a nefarious non-state organization that is just as genocidal as a real "government."

Earlier, I advanced the notion that if evolution looks like it has produced something improbable (for example, genocide) it's likely that something else has joined forces with it.

Having established the notion that genocide is a lot more common than we might have thought it was, and has been common since prehistoric times, yet is evolutionarily improbable, let's explore how it might have been acted out on the ancient African savannah. It was conducted not by governments, but by small tribal bands. First comes a dispute of some sort, a contest over hunting grounds, stands of important vegetation, deposits of obsidian or cosmetic pigments, foraging areas, pinch points on trade routes, possession of women (see "kleptogamy," below) and so forth. The fighters are able-bodied men. Non-combatants and women remain in the camp or hide in a redoubt. Some fighters are killed in the battle, and the losing side withdraws. The winning side may extract, or preserve, the resource of the contention, and also withdraw. But at some long ago time, a genetic innovation took place, due to an insult to a germ cell, say by a stray cosmic ray, a toxic compound from an ingested plant, too much sun, a pathogen, or the passage of time. The capacity for genocide arrives in a mutation in the genome of one person – most likely a male person – in one band.

It only needed to happen once. When the first male descendants of that mutated parent came of fighting age, they liked to keep on killing after winning the battle, and did not withdraw at the end of the combat period. They followed the losing fighters back to their camp, killed them, and there slaughtered the noncombatant civilians, principally the males and male children sired by those male adults. They neutered and enslaved some, if they believed they would be useful as workers. The point is that they didn't just remove the losing fighters from the gene pool as in conventional mortal combat; they removed from the gene pool ALL the males from the losing band, nonfighters as well as fighters, children as well as adult men. We will shortly discuss the fate of the females, which is quite different from that of the males.

In the prototypical modern genocide, the males of the losing side are mutilated and massacred wholesale by the males of the winning side. The Holy Bible [HB] documents, and celebrates, a great many genocides. A few samples: Deuteronomy 20: 10-17, wherein one may read the basic rules for conducting a genocide upon conquering a city; Numbers

31: 7-19, which repeats a detailed manifest of the fates of the captives after an Israelite victory over Midianites; Joshua chapters 6 through 12, which tell of several bloody massacres. We will revisit some of these citations when we discuss kleptogamy, below. Biblical passages are recent reports, in comparison to what went on in Africa in really ancient times. We are speculating, obviously, when we discuss what went on a million years ago. We weren't there, were we. Our genes were there, though, and they were also present in the armies of Joshua and other Biblical conquerors.

The massacres our species carries out today are very well documented in literature and in museums around the world. Because this activity is so crazy, I suspect it is not due to sequential, rational, non-instinctual thought but to relict instincts, instincts that were useful enough to be implanted in the genome much longer ago, instincts whose effects are horrid and embarrassing today, yet we seem to be stuck with them. The bibliography includes several reference works that I find difficult reading, for personal and familial reasons, e.g. [Morgenthau1919] and [Balakian1997]. Other genocide-related books I found gripping include [Guttman1993], [Kiernan2007] and [Power2002]. None of these mentions any connection to Darwinian evolution, and could be much improved thereby.

We'll now use our knowledge of what goes on today to model what went on a million years ago. It is plausible to do this, if we buy the argument that the innate capacity to commit genocide arose in prehistory, and magnified its presence as described here. If a few dozen of our more warlike modern people were placed on the ancient African savannah by a million-year time machine, this is very probably what they would do. There are still tribal people living in African forest and savannah today, and this IS what they do - see [Genocidewatch2012].

And imagine what would have happened if our speculation is correct about bloody deeds on the ancient African savannahs. The goal, and possibly the result, would be nothing less than the removal from the ongoing gene pool of the entire male population of each losing band. The gene pool from that point forward belongs to the band that just won, with its newfound heritable capacity for genocide. The next consequence is that they are right away ready to go out and conquer the next band, and the next. If it happens as described above, there would be no stopping it. It would spread very fast, fight by fight, band by band, generation by succeeding generation. This capacity, joined with that for mortal combat, could cover a continent in several generations. Large numbers of male fossils would - and do - show combat damage, even when only a few individuals had actually fought. Damage would be – and is – less in fossils of girls or young women. One is drawn to speculate that this behavioral eruption was what brought about the exodus and diaspora from Africa 100,000 to 250,000 years ago.

The Capacity for Kleptogamy

Females, as noted earlier, are the limiting resource for reproduction. The more women of childbearing age you have in your tribe, the faster it can grow, and the more rapidly it can replenish losses. A band that lives in African savannah and forest can increase its size – or prevent its decrease – in proportion to the number of fertile females in it. Since attacks from neighboring hostile bands may take place at random, bands need to reach a certain minimum size in order to repel raids and thereby remain viable. To this end, bands seek to increase their supply of women to the maximum number they can both control and defend from raids by other bands. Having too many women in one's band not only makes it difficult to control, but also invites attacks by raiding parties from bands that have excess males. The men of one band may trade for women with friendly neighbor bands, if they have anything to trade. Or, they may simply raid for them, taking them by force from weaker bands.

A vivid presentation of these general problems, along with a highly believable description of the necessary human coping mechanisms can be found in [Chagnon1968]. The situation Chagnon reports is very dynamic, especially when there are large numbers of bands of different compositions in contact with one another. All bands are under constant stress, day by day struggling to find game, choosing to raid and defend, to form or break alliances, to specialize in trade for one particular commodity, to gain women, and to be absorbed by or break away from a stronger nieghboring band. Treachery lurks in the shadows of every alliance.

I have chosen the word "kleptogamy" to describe the taking of women by one coalition from a foreign rival coalition, regardless of its circumstances and degree of force. The usual consequence of such a taking is a transfer of reproductive power, enhancing the spread of the male taker's genotype, and the corresponding dimunition of the genotype of the men from whom the women were taken. We could say this is its evolutionary purpose, if indeed evolution could be said to have a "purpose," but indeed it does not. It is easy, though, to see how a capacity for kleptogamy could be self-reinforcing. A band whose men did not practice kleptogamy would soon lose all of its women, and so perish wholesale, or be absorbed into a neighbor band as slaves. I know that in its dictionary definitions, the word "kleptogamy" is used in other ways, but I couldn't find a better word to use here – it does convey the general notion of taking women from a foreign coalition into one's own coalition.

There are lots of kinds of taking. Women in modern day wars might be taken peaceably, and assimilated as "war brides." The women may even initiate the taking, preferring to have their children fathered by one of the strong, victorious (and numerous) invaders than by their depleted countrymen, and raised in the victor's country rather than in the war-devastated country of their own birth. Women captured by force might be enslaved as "comfort women," and forced to provide sexual services for victorious soldiers at rest and recuperation. Women who remain alive during or after a massacre of males (see genocide, above) might be massively, formally, procedurally raped, as an exercise in national humiliation – the reminders of the humiliation continue throughout the adolescence and maturation of the unwanted offspring, and indeed throughout the lifetimes of the affected women ref: [Children of the Hated] in the Bibliography. Some kleptogamies, *e.g.* ones mentioned in [HB] make distinctions between virgins and nonvirgins, but I have no suggestions as to how that proclivity may have arisen. Two of the Biblical citations listed earlier in the discussion of genocide pay special attention to the treatment of the women during a massacre of the men, *i.e.* Deuteronomy 21:10-14 deals with the treatment of "comely" female captives; Deuteronomy 20:10-17 advises killing all the men but taking the women, dependents, cattle and other plunder; Numbers 31: lays out Moses' instructions on which women to take and which to kill with the men. To investigate Biblical kleptogamies further, search the Web for "holy bible kidnap women," and get half a million hits. No matter how it takes place, no matter how unpleasant the story, kleptogamy magnifies the reproductive opportunities of the winning males, and frees the species as a whole from any concern about inbreeding. The losing males having been killed or neutered, the only remaining male gametes in this mix come from the winning side. The male offspring, whether hated or not, will carry the genotype forward.

The Speed of it All

The accelerating effect of genocide and kleptogamy on their concomitant evolutionary processes should be clear by this point in the book. An ancient band of males who possess the five capacities, those for coalitionality, religion, mortal combat, genocide and kleptogamy, would spread their genotype with lightning speed, implanting it throughout the entire accessible prehistoric human world in a matter of months to decades, not centuries, and certainly not millenia.

What About Today's Atheists and Pacifists?

Once the five capacities began working together, there was no stopping the process short of letting it burn itself out. It rapidly spread into all contiguous habitable territory. Then the situation changed. The landscape was soon entirely occupied by bands that possessed all five capacities to greater or lesser degrees. The five capacities, once so decisive in combat, lost their effectiveness in determining the outcomes of battles. What brought about this change? After the general spreading had taken place, there were no more remaining target bands who lacked any or all of the capacities. A saturation level had been reached. An equilibrium took root. Raiding and attacking continued as before (as they still continue today). Genocide and kleptogamy continued as before (as they still continue today). But with the five capacities now balanced in the genomes of the combatants on both sides, none of the capacities had any further inroads to achieve. Each attacking band found itself confronting a defending band that was just as religious, just as combative, just as genocidal and just as kleptogamous as themselves, and had the home field advantage at that. With no more opportunities for the new innate capacities to turn the tide in battle, this chapter in the grand story of human evolution tapered to a close, with the capacities fully installed and well rooted in the genome at large. The capacities continue to exist today, and continue to wreak their devastation on us, but they are no longer being amplified.

Saturation stopped the process quite a bit short of 100%. How short? My favorite estimate for the capacity for religion, as reported above, is 80%, based on [Pew2012], which I interpret as revealing that 20% of humans lack the capacity for religion. Or if you prefer, you could accept 85% as implied by the "billion nonreligious people" claimed by [Epstein2004]. I have no quantitative estimates for the other capacities, and no surveys on which to base any. I invite readers to speculate on a link between pacifism and atheism, given that the capacities for combat and religion are tightly coupled. The saturation phenomenon is what results in the continuing presence of congenital atheists and pacifists in our population.

Reviewers have suggested that I need to repeatedly emphasize the fact that the fights I've been describing are not the battles of today's large armies, or even of Biblical nations, but raids carried out by small primitive bands. They took place hundreds of thousands of years ago, and did their evolutionary work at that time. Humans continue to conduct mortal combat, religion, genocide and kleptogamy today, but our present-day fights no longer amplify the five capacities as they did during distant times in Africa. Evolution, the great amplifier of lucky accidents, seems to have no way of removing them from our genome and psyche.

Summary of Chapter 6

From that day when the capacity for religion first arrived, on up to the present day and into the future, mortal combat continues, religion continues, genocide continues, and kleptogamy continues. What doesn't continue is their growth, the constant increase in the gene pool of the capacities that enable those behaviors. The capacity for religion has leveled out at about 80%. The gene pool is well mixed: replacement of one part with another part no longer makes any difference to the pool as a whole. These five traits have reached their maximum penetration. They will echo forever in our genome, but never claim the total population.

> "Say nothing of my religion. It is known to my God and myself alone. Its evidence before the world is to be sought in my life; if that has been honest and dutiful to society, the religion which has regulated it cannot be a bad one."
> - Thomas Jefferson

7. Writer's Apology

In this closing chapter, I have a few apologies to make, in at least two senses of the word.

Not an Argument For Or Against Religion

Some readers may feel that I have cast the capacity for religion, and by implication religion itself, in a bad light; another number may feel that I have unfairly impugned the beliefs of atheists. Agnostics could be miffed that I ignored them altogether. In my own defense, I repeat the first sentence of my introduction: "This is not an argument for or against religion."

I'm content with my belief system, and I am content with you having yours. I am content that you are content with yours. There's no need for yours and mine to be the same, but there is a need for each of us to tolerate the other's. Stores and libraries are filled with interesting books for and against individual religions and religion in general, but this isn't one of them. Given the innate nature of the capacity for religion, I don't think any of those for-and-against books will have any real effect on anyone.

If you still feel insulted, kindly consider that insult, like beauty, is in the eye of the beholder. So, if you feel that your cherished beliefs or disbeliefs have been demeaned, I sincerely apologize for providing a trigger for those feelings, but it is you, not I, whose mind provides the sparking flint and the scorching energy of that powder charge. As any competent therapist will tell you, those feelings are YOUR feelings. It is up to you to explore and manage them.

A Gory Mess, and No Exit – Sorry About That

Second, some critics have complained to me that I have dragged us into a gory mess, and failed to prescribe a way out of it. In an allied, but more encouraging vein, one of my most helpful reviewers tells me this should be a call for human change. I would feel honored if it were used to support such a call. It's definitely a start. But someone else will need to carry that torch. How about you? Indeed I have devoted large parts of my life to the cause of peace in the world. Presumably so have you. How are we doing these days? Anything new to report? This problem is similar in scope to the problem of slowing and reversing climate change. I can't imagine it will be easy.

For now, I have no universal evolution-based solution to propose. Have you? We need it. Don't hold it back. And about that gory mess: It was not I who dragged us into it. We were already in it up to our necks before we were born, indeed before our species moved out of Africa. I merely pointed it out, and identified the factors that were likely to have placed us in it. I did this by observing and connecting a few highly visible dots, and by telling an almost obvious story that almost anyone could have told.

It took our whole species hundreds of thousands or millions of years to drag itself into the gory mess, and it could take more than one man more than my remaining span of life to bury all the hatchets and lift us back out. You are now armed with fresh understanding of the problems of keeping peace and probity on the planet, an understanding based on evolution by natural selection. Solutions? I must leave that field to you, for now.

Shocked and Disappointed?

Third, the story may have simply shocked and disappointed you. I'm very sorry about that. You might have been expecting something nice or peaceful. Try as I might, I found no way to sugar-coat this tale of 'Nature, red in tooth and claw' (Tennyson, "In Memoriam, A.H.H." Canto 56). Now that I've told the story, I am sorry to have troubled you with it. Please know that I am not the only one who has explored and speculated on our violent origins and pondered their future.

I definitely did not invent the idea of the genetic heritability of our violent and holy instincts – that is old news. Indeed, our instincts have been there for all to see for as long as we have known about instincts. They have enabled us to kill and enslave fellow members of our own species through prehistory and into the present day.

Where else is this human horror discussed? For a quick experiment, try a Web search for "sons of Cain," and retrieve over eight million hits, many of them packed with serious vitriol. My writings contain no vitriol whatsoever: I shall gratefully accept your gratitude for that, should you offer it.

I admit the theory is a nasty one. If it is correct, and it probably is, it has seriously bloody implications for the future of humanity as it has had for the past. But your opinion matters. If what I've written here is wrong in whole or in part, you are encouraged (even obligated, by the conventions of scientific debate) to write your proposed corrections, or even a whole substitute theory. Kindly observe that this story, like all biological stories, is one of Darwinian evolution. Your corrections, opposing views and competing theories will be best received if they draw their strength and support from that same foundation.

My Proofs are Weak

Fourth, reviewers have complained to me that my facts are soft and proofs are therefore weak. No argument. In fact, they are soft and weak. Please swiftly lend a hand to repair their weaknesses. Some of my weaker arguments could seem so egregiously weak to you that they detract from the credibility of the stronger ones. Don't let that bother you. Reject the totally preposterously weakest ones and you may see that the whole story still holds itself together. Thomas Jefferson

[Jefferson1819] did something similar to his Bible with a pair of scissors, after all.

I have had to rely on suggestive heuristics instead of on actual "strong evidence." Nobody can yet cite strong evidence – for example of the heritability of the capacity for religion – because the chemical work has not yet been done, and the scanners and questionnaires are not yet equal to the task. I was pretty clear in the body of the writing that there are weaknesses of that sort. So, sorry about that, but it will be your job, not mine, to fill in the gaps. My time remaining on the planet is short, and I could not wait in the wings for the rest of you to conduct and publish hoped-for experimental results.

I am connecting important dots. The dots undeniably exist. I've connected them with lines that make sense to me. I accept that ironclad Euclidean proofs of the integrity of the figures made by the connected lines have not yet been constructed. Why did I rush into print?

Indeed almost all really good ideas roll out of the cradle before they can walk, don't they? *i.e.* ahead of the experimental evidence. Darwin, for example, did not know the mechanism of inheritance. It was obvious to him that a semi-dependable mechanism must exist, though: without such a mechanism, there would be no way for organisms to produce offspring that resembled their parents. He didn't know how it worked, but relied on it anyway because no better explanation of the experimental results had yet been placed on the table. His rousing good book [Darwin1859] is largely based on reasoning about observations of his own experiments in animal and plant domestication. He made good use of a great many specimens sent to him by friends all around the planet. Let's use that model here, as well.

In a truly special case of asserting the unproven, the existence of the "Higgs Boson" was postulated in 1964, but verified only after the passage of 48 years and the investment of many billions of euros. Did Higgs "rush into print?"

Newton didn't know how gravity works, and neither do today's physicists: all we know about gravity is what it does. Our engineers also have a precise and reliable understanding of what gravity does, and mostly we don't actually care how it works. We routinely make accurate predictive calculations for the behavior of systems in which the acceleration of gravity plays a role. Our buildings and bridges stand up. Our cannonballs arrive where they are aimed, possibly knocking down those very same buildings. Our rockets carry instruments to explore our cosmic system, with excellent accuracy. Can Isaac Newton, who gave us our quantitative law of universal gravitation without even a clue about how it works, be accused of rushing into print?

In the case of my current work – how evolution produced the religious mind – I only hope that the weaknesses in my writing will not cause it to be hammered into undeserved obscurity. I'd prefer to see those weaknesses stimulate the wider social efforts that eventually produce the needed research, and in consequence of that, the devoutly desired actual "strong evidence."

A proper method of handling our genetically modulated bad behaviors, now, that is something we need to discover. The labor will be well worth the investment, if we hope ever to contain the risks that these coevolved instincts expose us to.

It could happen faster than any of us expect. The young science of mol-bio leaps ahead with constantly surprising speed. I am writing this barely a decade after the first analysis of the human genome was published. I began the study just a decade or so before that historic event took place, and less than two decades after Kary Mullis invented the polymerase chain reaction, "PCR" – that made DNA sequencing possible. Thousands of other creatures have been sequenced since the human, and the process gets faster and cheaper every year.

Ordinary people have come to assume, perhaps without complete justification, that we are on the brink of using C-A-T-G sequences to unravel all sorts of mysteries about our common ancestry. I say you'd better not wait for me to look it up and cite it for you. You go do it.

At this point in the development of the theory about the five capacities, we do not need the jury to be unanimous, as in a criminal trial. All we need, as in a civil trial, is a preponderance of evidence.

How Can Anyone Assert That This is Good?

Fifth, some readers may erroneously conclude that I like these conclusions and their results, and think they are wonderfully good and beneficial things, evidence of God's Will and Man's perfection, and so forth.

It means nothing of the sort: Leibniz himself, famous (among his many other wonders) for asserting that ours is the best of all possible worlds, must have had to imagine some truly awful worlds in order to justify his use of the superlative "best." He was savagely satirized by Voltaire as the Dr. Pangloss in

"Candide." Like Pascal, he did his thinking in a century that produced the ruinous Thirty Years' War.

Evolution has no time to waste on perfecting things. It just provides enough to get by for a while under current conditions. As conditions change, it keeps moving on, leaving behind not only success and outright failure, but also a clutter of vestigial remains – legacies, bits of machinery that once worked OK but maybe aren't so pretty today. It works in a large, echo-filled space, in which its past choices may resonate for thousands or millions of years.

We live under greatly changed conditions, relative to those of our small, dispersed bands on the African savannah. The instinctual tools that were such powerful adaptations for us there and then, weigh us down and embarrass us here and now. Once, they were vital structural components, aiding us in our struggles against our conspecific enemies. Today, our conquest of the Earth is complete. We no longer need those components. But still their chemical memories remain in our DNA. We have no need to enter mortal combat against conspecific enemies anymore, but we still do it. We cope with it all the best we can.

We don't have to like something just because it's true and because it's part of us. The world offers many unlikable truths, and this story is just one of many such. These five capacities, acting in combination, don't seem to be very good for us or for the rest of our present day world, do they. Most of all, the fact that I've reported them does not mean that I expect you to like those instincts, either.

I do expect you to recognize that they are a part of us all (*i.e.* part of you and part of me, part of our ancestors, and part of our descendants) and to be conscious of them, and discuss them with your children. Possibly it will stimulate and augment your thinking about how to work against these deeply seated instincts that threaten our post-Enlightenment scientific progress and even the maintenance of our planetary home.

Certainly, anti-war efforts exist today. Hopeful and necessary, but limited in effect. Wars, with attendant rapes and genocides, continue to rage around our world. By reaching back to the root cause of the problem of war, *i.e.* to the evolution of these five capacities by natural selection, anti-war movements might strive to build on a richer and more accurately proven foundation, and find the fulcrum they so badly need to help them move the world. My prediction is that the eventual solutions, just like the problems, will be based on evolution by natural selection. After all, isn't every aspect of every process of life so based? Please get to work on these things.

A Bit Repetitious

Sixth, Some of you may complain that I occasionally repeat myself, presenting a given idea in a nearly identical way in two or more places. Some of that is intentional on my part. For one thing, I found it a struggle to present these five instinctive human capacities – those for coalitionality, mortal combat, religion, kleptogamy and genocide – as separate entities having separate actions and separate evolutionary careers.

As you have seen, the capacities themselves are coevolutionary. Rather than acting alone, they tend to show up with greatest presence while joined cooperatively in pairs, trios, quartets, and so forth. For another, these combinations come into play at different stages of our biological history, and their order of appearance is not entirely clear.

Every good evolution story, including that of our very own species, is a long, messy, tumbling, stumbling romp through discovery, disuse and rediscovery of adaptive innovations whose onset and development alternately speed up and slow down. If the book's apparent clutter and disorganization makes you struggle with the reading as I struggled with the writing, that's all to the good. One chaotic repeat begets another.

During my long-ago training to be a middle-school general-science teacher, I was taught more than once that some repetition helps the students in their comprehension efforts. Most readers are way past middle school, but ... how's it going? Perhaps in a subsequent edition I will be asked by the editors to make more of an effort to separate all the strands, present each one just once as a complete and isolated entity, tie them all together at the end and still have the story make sense. But don't wait for me, while you could be doing it yourself...

This Story is Very Nihilistic

Seventh, this picture that I present to you does indeed lay out causal relationships among some of the less pleasant innate characteristics of our species. It will take another book to describe how we also came to evolve the instinct that regards these characteristics as unpleasant, ugly and punishable. Meanwhile, the numerical count of the human population – the Darwinian measure of a species' success in a niche – continues to rise unimpeded. Few of us burden ourselves with thoughts of the long term. I offer here no value judgments of either trend.

No, This Isn't Group Selection

Eighth, no, these conjectures do not rely on the "Group Selection Hypothesis," (*q.v.*) that has been proposed for certain classes and orders of animals (*hymenoptera*, in particular). I find no fault with that theory where it helps explain something that individual selection can't. Feel free to try, if you like; I don't base any of my arguments on it, though. What I've described is basic Darwinian selection. It acts at the level of the individual. We are individuals, and we act as individuals. Our genotypes are the genotypes of individuals, no two alike. We fight, win and lose as individuals. When we team up to accomplish something too great for one individual to accomplish alone, our team is led by an individual. Our coalitionality is fluid, and our alliances so inconstant as to prevent "kin altruism" from being the dominant force in holding our groups together. We can rarely summon a mutually related brood for anything more lasting than a family reunion.

The engine that most powerfully amplifies and attenuates the alleles for four of the five traits described here – and possibly the fifth as well – is intraspecific hand to hand mortal combat. It is a quick fight to the death between two individual men of the same species. One man's genes are culled out, and the other man's (only slightly different) genes endure. In the prehistoric times when all this was being sorted out, proto-humans fought hand to hand, and had no way of acting against an entire opposing population. They did not possess weapons of mass destruction.

Even as "advanced" as we are today, it is difficult to find a human coalition (team, union, band, choir, company, troop, *etc.*) that can act as though it had a single mind. We are sympathetic to our leaders and agreeable to the mission statements and goals of our coalitions, but we are not instinctively capable of acting as a group, *i.e.* as a unified mass of genetically identical organisms. Such a mass should not require assignment of a few members to form a leadership cadre or a management structure. There would be no need for business schools, or for any kind of management education. Teams of humans cannot unify even a small portion of their thoughts and actions without a working leadership structure.

So, Group Selection theory may work in some places, but I have not relied on it here.

Is Any Follow-Up Planned?

Ninth, yes, time and health permitting. As noted somewhere above, this project (some called it "George's Dangerous Idea.")

has been on low boil for parts of three decades. That period produced not only this slim and economical Volume I, but also an accumulation of useful and interesting reference material: some of it appears as citations in the Bibliography or in the text. A planned Volume II (which could possibly stretch into Volume III) is in preparation. It/they will contain excerpts from works selected from the Bibliography of Volume I, plus commentaries from the author responding specifically to those authors and those excerpts. Harder to manage is the fluid collection of several thousand Web links to which a visit might be re-edifying. I hope to be able to post them all in one place where interested readers can click them directly. I might be able to do that in an e-book edition, or maybe on the publisher's web site (amazon.com). Further, I plan to save any correspondence that I find interesting, and reply publicly to any that I find exceptionally interesting. Perhaps I will be allowed to provide a public comment board so that readers can contact each other directly? A message board might be both entertaining and useful. I'm sure you will let me know somehow.

So, as you see, it's just beginning. If you enjoyed this book, please consider leaving a kind review on the Amazon site next to the book.

The Dissent of Man

Tenth (and finally), critics will be pleased that I leave room for earnest dissent. My wife earnestly requests me to leave the room before the dissent becomes too earnest.

G.V., California, 2015

Bibliography

[Ackroyd1997]
Ackroyd, Peter
Blake: A Biography
Random House
ISBN 0-7493-9176-6 or 9780749391768

[Allen1990]
Allen, Steve
Steve Allen on the Bible, Religion and Morality
Prometheus Books, Buffalo NY 1990
ISBN 0-87975-638-1

[Alper2006]
Alper, Matthew
The God Part of the Brain
Sourcebooks Inc. 2006
ISBN 978-1-4022-0748-8

[Ansary2009]
Ansary, Tamim
Destiny Disrupted
PublicAffairs / Perseus Books Group
ISBN-13: 978-1-58648-606-8

[Ardrey1961]
Ardrey, Robert
African Genesis
A Delta Book (Dell, New York)
Library of Congress Cat. No. 61-15889

[Atkinson2007]
Atkinson, Rick
The Day of Battle: The War in Sicily and Italy 1943-1944
Henry Holt & Co. 2007
ISBN-13   978-08050-8861-8

[Atran2002]
Atran, Scott
In Gods We Trust: The Evolutionary Landscape of Religion
Oxford University Press 2002
ISBN 0-19-514930-0

[Avalos2005]

Avalos, Hector
Fighting Words (The origins of religious violence)
Prometheus Books 2005
ISBN 1-59102-284-3

[Balakian1997]
Balakian, Peter
Black Dog of Fate
HarperCollins Basic Books 1997
ISBN 0-465-000704-X

[Barzun2000]
Barzun, Jacques
From Dawn To Decadence, 1500 to the Present
HarperCollins 2000
ISBN 0-06-092883-2 (pbk)

[Bellah2011]
Bellah, Robert Neelly
Religion in Human Evolution: from the Paleolithic to the Axial Age
Belknap/Harvard 2011
ISBN 978-0-674-06143-9

[Boorstin1983]
Boorstin, Daniel J.
The Discoverers: A History of Man's Search to Know His World and Himself
Random House, 1983, Vintage, 1985
ISBN 0-394-72625-1 (pbk)

[Brockman2007]
Brockman, John (Editor)
What is YOUR Dangerous Idea? – today's leading thinkers on the unthinkable
HarperCollins, New York 2007
ISBN 978-0-06-121495-0

[Browne1643]
Browne, Sir Thomas
Religio Medici **(The Religion of a Doctor)**
(Frank L. Huntley, Ed.)
Everyman's Library (E.P.Dutton), New York, 1951

[Bryson2003]
Bryson, Bill
A Short History of Nearly Everything
Broadway Books / Random House 2003
ISBN 0-7679-0817-1

[Bryson2010]
Bryson, Bill
At Home: A Short History of Private Life
Anchor Books / Random House 2011
ISBN 978-0-7679-1939-5

[Buruma1994]
Buruma, Ian
The Wages of Guilt: Memories of Wan in Germany and Japan
Farrar Straus Giroux 1994
ISBN 0-374-28595-0

[Buss1994]
Buss, David M.
The Evolution of Desire: strategies of human mating
BasicBooks/HarperCollins 1994
ISBN 0-465-07750-1

[Buss2005]
Buss, David M. (Ed.)
The Handbook of Evolutionary Psychology
Wiley 2005
ISBN 978-0471264033

[Butcher2011]
Butcher, Tim
Chasing the Devil: A Journey Through Sub-Saharan Africa in the Footsteps of Graham Greene
Atlas & Co, Publishers, New York.
Distributed to the Trade by W.W. Norton & Co
ISBN 978-1-935633-29-7

[Carroll2009]
Carroll, Sean B.
Remarkable Creatures: Epic adventures in the search for the origins of species
Mariner /Houghton Mifflin Harcourt 2009
ISBN 978-0-15-101485-9

[Chagnon1968]
Chagnon, Napoleon A.
Yanomamö – the Fierce People
Holt, Rinehart and Winston 1968
ISBN 0-03-071070-7

[Cheney2007]
Cheney, Dorothy L. and Seyfarth, Robert M.
Baboon Metaphysics – The Evolution of a Social Mind
University of Chicago Press, 2008
ISBN978-0-226-10244-3

[Clark1969]
Clark, W.E. Le Gros
The Fossil Evidence for Human Evolution
University of Chicago Press, 2nd Edition 1964, 1969
Library of Congress Cat. Card # 64-22250

[Constitution]
Constitution of the United States

[Cottrell1962]
Cottrell, Harold
Wonders Of The World
Tempo Books / Holt, Rinehart, Winston 1962
Libr. Of Congress Card # 59-6565

[Damasio1994]
Damasio, Antonio R.
Descartes' Error: Emotion, Reason and the Human Brain
Penguin/Putnam 2005
ISBN 0 14 30.3622 X (Pbk.)

[Darwin1845]
Darwin, Charles
The Voyage of the Beagle (1845)
And
[Darwin1859]
On the Origin of Species (1859)
Both Included in
From So Simple a Beginning-the four great books of Charles Darwin
Edward O. Wilson, editor, 2006
W.W. Norton, New York, 2006
ISBN 0-393-06134-5 (hardcover)

[Dawkins1976]
Dawkins, Richard
The Selfish Gene
Oxford University Press
30th Anniversary Edition, 2006
ISBN 0-19-929115-2 (pbk.)

[Dawkins2003]
Dawkins, Richard
A Devil's Chaplain: reflections on hope, lies, science and love
Houghton Mifflin, New York 2003, Mariner Edition 2004
ISBN 0-618-48539-2 (pbk)

[Dawkins2006]
Dawkins, Richard
The God Delusion
Houghton Mifflin New York 2006
ISBN 978-0-618-68000-9

[Dennett1995]
Dennett, Daniel C.
Darwin's Dangerous Idea
Simon and Schuster Paperbacks, New York 1995
ISBN 978-0-684-82471-0

[Dennett2006]
Dennett, Daniel C.
Breaking the Spell: Religion as a Natural Phenomenon
Viking (Penguin) 2006
ISBN 0-670-03472-X

[Diamond1999]
Diamond, Jared
Guns, Germs and Steel: The Fates of Human Societies
Norton, 1999
ISBN 978-0-393-31755-8

[Dobzhansky1973]
Dobzhansky, T.
"Nothing in Biology Makes Sense Except in the Light of Evolution"
Essay in "American Biology Teacher" March 1973
Posted on Internet at

[Durkheim1912]
Durkheim, Emile
The Elementary Forms of Religious Life
Translated by Carol Cosman, 2001
Oxford University Press
ISBN 978-0-19-954012-9

[Dyson2012]
Dyson, George
Turing's Cathedral
Vintage/Random House 2012
ISBN 978-1-4000-7599-7

[Eakin1975]
Eakin, Richard
Great Scientists Speak Again
University of California Press 1975
ISBN 0-520-03087-7 (paper)

[Ehrman2005]
Ehrman, Bart D.
Misquoting Jesus: The story of who changed the bible, and why
HarperOne, an imprint of HarperCollins Publishers New York 2005
ISBN 978-0-06-085951-0

[Einstein1979]
Einstein, Albert
Autobiographical Notes
Open Court Publishing Co. 1979

[Epstein2004]
Epstein, Greg M.
Good Without God – What a Billion Nonreligious People *do* Believe
HarperCollins 2004
ISBN 978-0-06-167011-4

[Fadiman1988]
Fadiman, Clifton
The Lifetime Reading Plan
Third Edition, Perennial Library/Harper&Row 1988
ISBN 0-06-055066-X

[Frazer1922]
Frazer, Sir James George, F.R.S., F.B.A.,
The Golden Bough – a study in magic and religion
(1-volume abridged edition) MacMillan 1951

[FFRF]
"Freethought Today"
Periodical newspaper of the Freedom from Religion Foundation

[Friedman2003]
Friedman, Thomas L.
Longitudes and Attitudes: The World in the Age of Terrorism
Anchor Books 2003
ISBN 1-4000-31257-7 (pbk.)

[Gaarder1996]
Gaarder, Jostein
Sophie's World: a novel about the history of philosophy
Berkley / Farrar, Straus & Giroux, 1996
ISBN 0-425-15225-1

[Gardner1957]
Gardner, Martin (Ed.)
Great Essays in Science
Contains several relevant essays, including
Eddington, Sir Arthur: "The Decline of Determinism Ch. 4"
from **New Pathways in Science, Ch. 4** (Cambridge University Press)
Whitehead, Alfred North: "Religion and Science"
from **Science and the Modern World Chapter 12** (Cambridge University Press)
Washington Square Press, New York 1957

[Gay1966]
Gay, Peter
The Enlightenment: an Interpretation: The Rise of Modern Paganism
Norton 1977
ISBN 0-393-31302-6

[Genet2007]
Genet, Russell Merle
Humanity: The Chimpanzees who Would be Ants
Collins Foundation Press, Santa Margarita CA 93453
ISBN 0-9788441-0-6

[Gilgamesh]
Many editions and commentaries of this ancient work exist, for example Sandars, N.K.
The Epic of Gilgamesh
Penguin 1972: ISBN 978-0-14-044100-0

[GftAoP1976]
Group for the Advancement of Psychiatry
Mysticism: Spiritual Quest or Psychic Disorder
Publ. No. 97 Volume IX
Group for the Advancement of Psychiatry, Inc. New York 1976
ISBN 87318-134-4

[Green1982]
Green, Ruth Hurmence
The Born-Again Skeptic's Guide to the Bible, with the Book of Ruth
Freedom From Religion Foundation, Madison Wisconsin, 1982
ISBN 1-877733-01-6

[Gould2006]
Gould, Stephen Jay (Steven Rose and Paul McGarr eds., fwd by Oliver Sacks)
The Richness of Life: The Essential Stephen Jay Gould (essays spanning SJG's career)
W.W. Norton 2006
ISBN 978-0-393-06498-8

[Graves1960]
Graves, Robert
The Greek Myths Vols. 1 and 2
Penguin 1960
ISBN 0-14 – 001026-2 and 0-14 – 001027-2

[Guilaine2005]
Guilaine, Jean and Zammit, Jean (Hersey, M. transl. From French)
The Origins of War : Violence in Prehistory
Blackwell Publishing 2005
ISBN 1-4051-1260-3

[Gutman1993]
Gutman, Roy
A Witness to Genocide: The 1993 Pulitzer Prize-Winning Dispatches on the "Ethnic Cleansing" of Bosnia
"A Lisa Drew Book"
MacMillan 1993
ISBN 0-02-546750-6
ISBN 0-02-032995-4 (pbk.)

[Gwynne2010]
Gwynne, S. C.
Empire of the Summer Moon: Quanah Parker and the Rise and Fall of the Comanches, the Most Powerful Indian Tribe in American History
Scribner 2011
ISBN 978-1-4165-9106-1 (pbk.)

[Hall1992]
Hall, Zach W.
An Introduction to Molecular Neurobiology
Sinauer Associates Inc. Sunderland MA 01375 USA
ISBN 0-87893-307-7

[Hamer2004]
Hamer, Dean
The God Gene
Anchor Books, 2004
ISBN 0-385-72031-9

[Hamilton1942]
Hamilton, Edith
Mythology: Timeless Tales of Gods and Heroes
Mentor/Dutton Signet/Penguin 1969 pbk
(Reprint of hardback by Little, Brown)
ISBN 0-451-62803-9

[Harris2004]
Harris, Sam
The End of Faith
W.W. Norton & Company, New York 2004
ISBN 978-0-393-32765-6 (paperback)

[Hawley2001]
Hawley, Jack
The Bhagavad-Gita: a Walkthrough for Westerners
New World Library, Novato CA 2001
ISBN 1-57731-147-7

[HB]
Holy Bible
Various authors.
Many editions of reasonable consistency and agreement exist, *e.g.*
The New English Bible
Various editors
Oxford University Press, Cambridge University Press 1970

[Hersch2010]
Hersch, Karen K.
The Roman Wedding: Ritual and Meaning in Antiquity
Cambridge University Press, 2010   ISBN 978-0-521-12427-0 (pbk.)

[Hitchens2007-1]
Hitchens, Christopher
God is not Great: How Religion Poisons Everything
Twelve/Grand Central/Hachette, New York 2007
ISBN 978-0-446-57980-3

[Hitchens2007-2]
Hitchens, Christopher, editor
The Portable Atheist
Da Capo / Perseus 2007
ISBN-13: 978-0-306-81608-6

[Hobbes 1651]
Hobbes, Thomas
Leviathan, or the Matter Forme and Power ...
Widely available *e.g.* ISBN 9781463649937

[Horgan2003]
Horgan, John
Rational Mysticism – Dispatches from the border between Science and Mysticism
Houghton Mifflin 2003
ISBN 0-618-06027-8

[Huberman2007]
Huberman, Jack
The Quotable Atheist
Nation Books (Avalon), New York 2007
ISBN 978-1-56025-969-5

[Hunter1952]
Hunter, John A.
Hunter
Safari Press (1999) Long Beach Calif.
ISBN 978-1-57157-243-1

[Jacoby2013]
Jacoby, Susan
The Great Agnostic: Robert Ingersoll and American Freethought
Yale University Press 2013
ISBN 978-0-300-13725-5

[James1910]
James, William
The Varieties of Religious Experience
Included in **Writings 1902-1910**
Library Classics 1987
ISBN 0-940450-38-0

[James1896]
James, William
The Will To Believe
Bound with **Human Immortality** Dover 1956
ISBN 486-20291-7

[Jefferson1819]
Jefferson, Thomas
The Life and Morals of Jesus of Nazareth
(also known as **The Jefferson Bible**)
Various editors, commentators, publishers *e.g.*
N.D. Thompson 1902
http://uuhouston.org/files/The_Jefferson_Bible.pdf

[Johnsen2013]
Johnsen, Gregory D.
The Last Refuge
W.W. Norton, 2013
ISBN 978-0-393-08242-5 (pbk.)

[Johnson1991]
Johnson, Haynes
Sleepwalking Through History: America in the Reagan Years
W.W. Norton 1991
ISBN 0-393-02937-9

[Kandel1991]
Kandel, Eric R. with Schwartz, J. and Jessell, T.
Principles of Neural Science (3rd Edition)
Elsevier 1991
ISBN 0-444-01562-0

[Kandel2006]
Kandel, Eric R.
In Search of Memory: the Emergence of a New Science of Mind
Norton Paperback 2007
ISBN 978-0-393-32937-7-pbk

[Katz2000]
Katz, Leonard D. Editor
Evolutionary Origins of Morality –cross disciplinary perspectives
Imprint Academic, Bowling Green Ohio 2000
ISBN   0 907845 07 X (paperback)

[Keeley1996]
Keeley, Lawrence H.
War Before Civilization – The Myth of the Peaceful Savage
Oxford University Press 1996
ISBN-13 978-0 – 511912-1 (Pbk.)

[Kenneally2008]
Kenneally, Christine
The First Word: the Search for the Origins of Language
Viking-Penguin 2008
ISBN 978-0-14-311374-4 (pbk)

[Kiernan2007]
Kiernan, Ben
Blood and Soil: A World History of Genocide and Extermination from Sparta to Darfur
Yale University Press 2007
ISBN 978-0-300-10098-3

[Kindlon2000]
Kindlon, Dan and Thompson, Michael with Barker, Teresa
Raising Cain: Protecting the Emotional Life of Boys
Ballantine/Random House, New York ad Toronto 2000
ISBN 0-345-43485-4

[Langer1952]
Langer, William L. (editor)
An Encyclopedia of World History
Houghton Mifflin / Riverside Press 1952
Lib. Of Congress Cat. No 52-9589

[Lattimore1967]
The Odyssey of Homer
Harper&Row 1977
ISBN 0-06-012531-4

[Lewis1942]
Lewis, C.S.
The Screwtape Letters
Harpercollins Paperback (2001)
ISBN 978-0-06-065293-7

[Linden2007]
Linden, David
The Accidental Mind
Belknap/ Harvard 2007
ISBN-10: 0674024788
ISBN-13: 978 0674024786

[Linden2011]
Linden, David J.
Pleasure: How Our Brains Make Junk Food, Exercise, Marijuana, Generosity, and Gambling Feel So Good
ONEWorld Publications, 2011
ISBN-10: 1851688242
ISBN-13: 978-1851688241

[Lodish1995]
Molecular Cell Biology
Lodish, Baltimore *et al*. 3rd edition
Scientific American Books / W.H. Freeman
ISBN0-7167-2380-8

[Lorenz1966]
Lorenz, Konrad
On Aggression (Transl. by the author of *Das Sogenannte Boese*)
Houghton Mifflin Harcourt 1966
ISBN 978-0-15-668741-6

[Macchiavelli1999]
Macchiavelli, Niccolo
The Prince
Translated by Ricci, L., Revised by Vincent, E.R.P.
Signet Classics, / Penguin Putnam 1999
ISBN 0-451-52746-1

[Macrone1993]
Macrone, Michael
Brush up your Bible!
Cader Books, New York 1993
ISBN 0-06-270024-3

[Manchester 1993]
Manchester, William
A World Lit Only By Fire: The Medieval Mind and the Renaissance; Portrait of an Age
Back Bay Books (Little, Brown) 1993
ISBN 0-316-54556-2 (pbk.)

[Mann2006]
Mann, Charles C.
1491: New Revelations of the Americas Before Columbus
Vintage/Random House 2006
ISBN 1-4000-3205-1, 978-1-4000-3205-1

[Maraniss2003]
Maraniss, David
They Marched Into Sunlight – War and Peace in Vietnam and America, October 1967
Simon and Schuster 2003
ISBN 0-7432-6104-6

[Mazzetti2013]
Mazzetti, Mark
The Way of the Knife: The CIA, a Secret Army, and a War at the Ends of the Earth

The Penguin Press, New York 2013
ISBN 978-1-59420-480-7

[McDannell1990]
McDannell, Colleen and Lang, Bernard
Heaven – a History
Random House/ Vintage Books, 1988, 1990
ISBN 0-679-72520-2

[Morgenthau1919]
Morgenthau, Henry
Ambassador Morgenthau's Story The documented account of the Armenian Genocide
New Age Publishers, Plandome New York
(Reprint of 1919 Doubleday edition)

[Moorehead1960]
Moorehead, Alan
The White Nile
Dell Publishing Co., Inc. First Dell Printing March 1962 (pre-ISBN)

[Moorehead1969]
Moorehead, Alan
Darwin and the Beagle
Penguin Books / Harper & Row 1969
ISBN 0-14-003327-0

[Najarian1986]
Daughters of Memory
City Miner Books / McNaughton and Gunn, printers
ISBN 0-933944-13-6

[NCSE2008]
National Committee for Science Education, Inc.
Voices for Evolution
P.O. Box 9477, Berkeley CA94709
ISBN 978-0-6152-0461-1

[NCSE2010]
National Center for Science Education, Inc.
NCSE Reports Vol. 30 No.3 May-Jun 2010
P.O. Box 9477, Berkeley CA94709
ISBN 1064-2358

[Newberg2002]
Newberg A., D'Aquili E. and Rause V.
Why God Won't Go Away: Brain Science and the Biology of Belief
Ballantine (Random House) 2002
ISBN 0-345-44034-X

[Newberg2006]
Newberg, Andrew, M.D. with Waldman, Mark Robert
Born to Believe – God, Science and the origin of ordinary and extraordinary beliefs or **Why we Believe What we Believe**
Free Press /Simon and Schuster
ISBN 978-0-7432-7498-2

[Pagels1989]
Pagels, Elaine
The Gnostic Gospels
Vintage, 1989
ISBN 0-679-72453-2 (pbk.)

[Pagels1995]
Pagels, Elaine
The Origin of Satan
Random House, 1995
ISBN 0-679-40140-7

[Paine2004]
Paine, Thomas
The Age of Reason
Dover 2004 (unabridged replication of Putnam, 1896)
ISBN 0-48643393-5 (pbk.)

[Peitgen1986]
Peitgen, Hans-Otto and Richter, Peter H.
The Beauty of Fractals: Images of Complex Dynamical Systems
Springer-Verlag, 1986
ISBN-13: 978-3540158516
ISBN-10: 3540158510

[Peters1976]
Peters, Alan with Palay and Webster
The Fine Structure of the Nervous System: The Neurons and Supporting Cells
W. B. Saunders Company 1976

Philadelphia PA 19105 USA
ISBN 0-7216-7207-8

[Pinker1997]
Pinker, Steven
How the Mind Works
Norton 1997
Penguin Group 1999
ISBN 978-0-140-24491-5

[Pinker2002]
Pinker, Steven
The Blank Slate – the modern denial of human nature
Penguin Group 2002
ISBN 0 14 20.0334 4 (pbk.)

[Power2002]
Power, Samantha
A Problem from Hell: America and the Age of Genocide
Basic Books / Perseus – 2002, "A New Republic Book"
ISBN 0-465-06150-8

[Quirk2006]
Quirk, Joe
It's Not You, it's Biology: the Science of Love, Sex and Relationships
Alternate title **Sperm are from Men, Eggs are from Women**
Running Press Book Publishers, Philadelphia 2006
ISBN 978-0-7624-3526-1

[Richter1986]
see [Peitgen1986]

[Rosenberg1971]
Cell and Molecular Biology – an appreciation
Holt, Rinehart and Winston 1971
ISBN 0-03-085312-5

[Russell 1963]
Russell, Bertrand
Mysticism and Logic
Unwin Books 1963

[Sacks2012]
Sacks, Oliver
Hallucinations
Knopf 2012
ISBN 978-0-307-95724-5

[Sagan1995]
Sagan, Carl
The Demon-Haunted World: Science as a Candle in the Dark
Random House 1995
ISBN 0-394-53512-X

[Shermer2004]
Shermer, Michael
The Science of Good and Evil
Henry Holt / Times Books 2004
ISBN 0-8050-7520-8

[Shlain1998]
Shlain, Leonard
The Alphabet Versus the Goddess: The conflict between Word and Image
Arkana/Penguin/Putnam/Viking 1998
ISBN 0 14 01.9601 3

[Smith2008]
Smith, Alexander McCall
The Good Husband of Zebra Drive
Anchor Books/Random House 2008
ISBN 978-1-4000-7572-0

[Smits2010]
Smits, Rik
The Puzzle of Left-Handedness
Reaktion Books Ltd., London UK
ISBN 978-1-178023-043-6

[Stern2003]
Stern, Jessica
Terror in the Name of God: Why Religious Militants Kill
ECCO / HarperCollins 2003
ISBN 0-06-050532-8 (pbk.)

[Storr1996]
Storr, Anthony
Feet of Clay: a Study of Gurus
HarperCollins 1996
ISBN 000 255563 8

[Thayer 2004]
Thayer, Bradley A.
Darwin and International Relations
University Press of Kentucky 2004
ISBN 0-8131-2321-6

[Turnbull1961]
Turnbull, Colin
The Forest People
The Reprint Society of London, Ltd. / World Books 1963

[Turner1993]
Turner, Alice K.
The History of Hell
Harcourt Brace 1993
ISBN 0-15-140934-X

[Valiant2013]
Valiant, Leslie
Probably Approximately Correct
Nature's Algorithms for Learning and Prospering in a Complex World
Basic Books / Perseus 2013
ISBN 978-0-465-03271-6

[Vonnegut1963]
Vonnegut, Kurt, Jr.
Cat's Cradle
"Delta" label of Delacorte Press/Seymour Lawrence
ISBN 0-385-33348-X

[Wade2009]
The Faith Instinct: How religion evolved and why it endures
Penguin Press 2009
ISBN978-1-59420-228-5

[Webster2008]

Webster's New World Medical Dictionary
Third Edition, Wiley 2008
ISBN-10: 0470189282 , -13: 978-0470189283

[West1998]
West, Thomas G. and Grace Starry (translators/commentators)
Four Texts on Socrates: Plato's Euthyphro, Apology and Crito, Aristophanes' Clouds
Cornell University Press 1998
ISBN 0-8014-8574-6

[White2012]
White, Matthew
The Great Big Book of Horrible Things: the Definitive Chronicle of History's 100 Worst Atrocities
W.W. Norton & Co., 2011
ISBN 978-0-393-08192-3

[Will1990]
Will, George F.
Men at Work: the Craft of Baseball
Harper Perennial / HarperCollins 1990
ISBN 0-06-097372-2

[Williams1997]
Williams, George C.
The Pony Fish's Glow – and other clues to plan and purpose in nature
HarperCollins / Basic Books 1997
ISBN 0-465-07281-X

[Williams2002]
Williams, Paul L.
The Complete Idiot's Guide to the Crusades
Alpha Books / Penguin Group 2002
ISBN 0-02-864243-0

[WilsonD2002]
Wilson, David Sloan
Darwin's Cathedral – evolution, religion and the nature of society
University of Chicago Press 2002
ISBN 0-226-90134-3

[Winokur1987]
Winokur, Jon

The Portable Curmudgeon
New American Library Penguin 1987
ISBN 0-453-00565-9

[Winsor1956]
Winsor, Frederick and Parry, Marian
The Space Child's Mother Goose
Simon and Schuster, 1956 or 1958
Possible reissue 2001 by Purple House Press ISBN 9781930900073

[Wright1994]
Wright, Robert
The Moral Animal – Why we are the way we are: The New Science of Evolutionary Psychology
Vintage Books 1995
ISBN 0-679-40773-1

[Wright2009]
Wright, Robert
The Evolution of God
First Edition Little, Brown / Hachette June 2009
ISBN 978-0-316-73491-2 (hc)

Part 2: Video Resources

[AAI2007]
Richard Dawkins Foundation for Reason and Science
Atheist Alliance International 2007 Convention
(two_dvd set, 552 minutes)
Several speakers: Dawkins, Harris, Hitchens, Dennett, Hirsi Ali, Thomson, Scott, Chapman, Tabash
www.RichardDawkinsFoundation.org 2007

[FourHorsemen2007]
Richard Dawkins Foundation for Reason and Science
The Four Horsemen Episode One
(one DVD, 120 min)
Several Speakers: Dawkins, Dennett, Harris, Hitchens
www.RichardDawkinsFoundation.org 2007

[Flemming2005]
Flemming, Brian *et al.*
The God who Wasn't There: a Film Beyond Belief
(one dvd 259 mins)
Talks by Butcher, Carrier, Dundes, Harris, Mikkelson, Price, Sipus
Beyond Belief Media, www.thegodmovie.com 2005

[DawkinsDVD2006]
Dawkins, Richard
Root of All Evil?
(two dvd's 160 min)
www.RichardDawkins.net and www.Skeptic.com 2006

[HC1]
History Channel
Archenemy-the Philistines
Documentary AAE-71914
From the series "Mysteries of the Bible"
www.historychannel.com

[HC2]
History Channel
Battles, B.C. the complete Season One,
Documentary AAAE154711
www.historychannel.com

[HC3]
History Channel
Bible Battles – history's bloodiest battles …
Documentary AAAE77622
www.historychannel.com

[Miller2008]
Miller, Jonathan
The Atheist Tapes
(two dvd's, 6 interviews, 180 minutes)
Interviews with Dawkins, Dennett, McGinn, Miller, Turner, Weinberg
http://www.secularphilosophy.com or www.ffrf.org

[TWM2001]
Several authors
Genocide: The Horror Continues
L4811DVD
Part of **The Genocide Factor** (a four-DVD set)
Media Entertainment Inc., 2001
www.tmwmedia.com

[Children of the Hated]
Children of the Hated
http://www.rnw.nl/english/radioshow/children-hated

Internet and Periodical Resources

[Bower2009]
Bower, Bruce
"Tracking the Inner World of Suspicion"
(Web edition is titled "The Inner Worlds of Conspiracy Believers")
Science News June 20 2009 Vol.175 #13 P.11

[Hauser2009]
Hauser, Marc
"Origin Of The Mind"
Scientific American, Sept. 2009 pp 44ff

[Hoover2011]
Hoover, Kelli *et al.*
"A Gene for an Extended Phenotype"
Science 9 September 2011: 333 (6048), 1401.

[Kolbert2011]
Kolbert, Elizabeth
"Annals of Evolution: Sleeping with the Enemy -- What happened between Neanderthals and us?"
The New Yorker ISSN0028792X
Vol. LXXXVII (87) No. 24, August 15 & 22, 2011

[McAuliffe2012]
McAuliffe, Kathleen
"How Your Cat is Making You Crazy"
Atlantic Monthly, March 2012

[Moyer2009]
Moyer, Michael
"Origins: Religious Thought"
Scientific American, Sept. 2009 p.92

[Sacks2010]
Sacks, Oliver
"Face-Blind" (A Neurologist's Notebook)
The New Yorker, August 30, 2010 page 36ff.

[Joseph2010]
Joseph, Pat
"Under Discussion: The (Really) Big Picture"
Interview with Walter Alvarez

California Magazine
Fall 2009 pp39-44

[Zimmer2014]
Zimmer, Carl
"Meet Nature's Nightmare Mindsuckers"
National Geographic November 2014 Pages 36ff.

[Wikipedia:Irreligion]
"Irreligion by Country"
http://en.wikipedia.org/wiki/Irreligion_by_country

[Pew2012]
PewResearch
Religion & Public Life Project 2012
http://www.pewforum.org/2012/

[Genocidewatch2012]
Genocide Watch
http://www.genocidewatch.org/alerts/countriesatrisk2012.html

[Levinger2006]
Levinger, Matt
"Genocide: Lessons from the 20th Century"
United States Holocaust Memorial Museum
April 1, 2006
http://tinyurl.com/pnxfcxf

[Mayell2003]
Mayell, Hillary
"Genghis Khan a Prolific Lover, DNA Data Implies"
National Geographic News, Feb. 14, 2003
http://tinyurl.com/637t

www.ingramcontent.com/pod-product-compliance
Lightning Source LLC
Chambersburg PA
CBHW071420170526
45165CB00001B/336